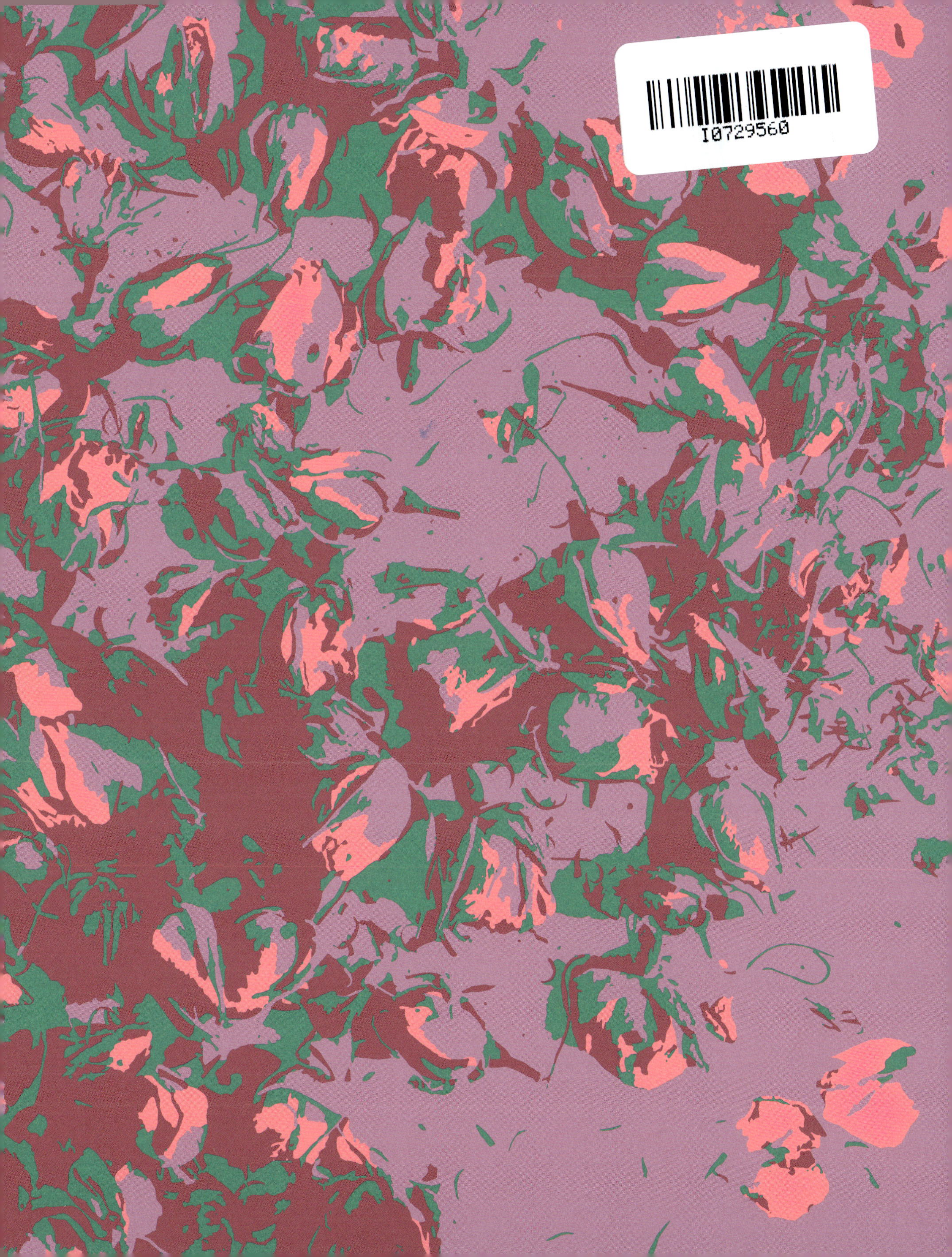

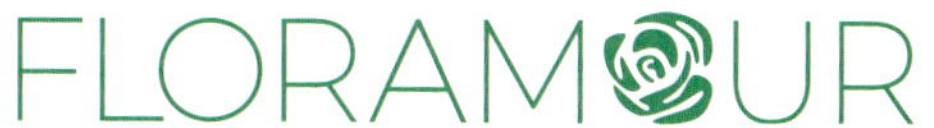

ANJA KLAFFENBACH

ROSES
ROSEN

teNeues

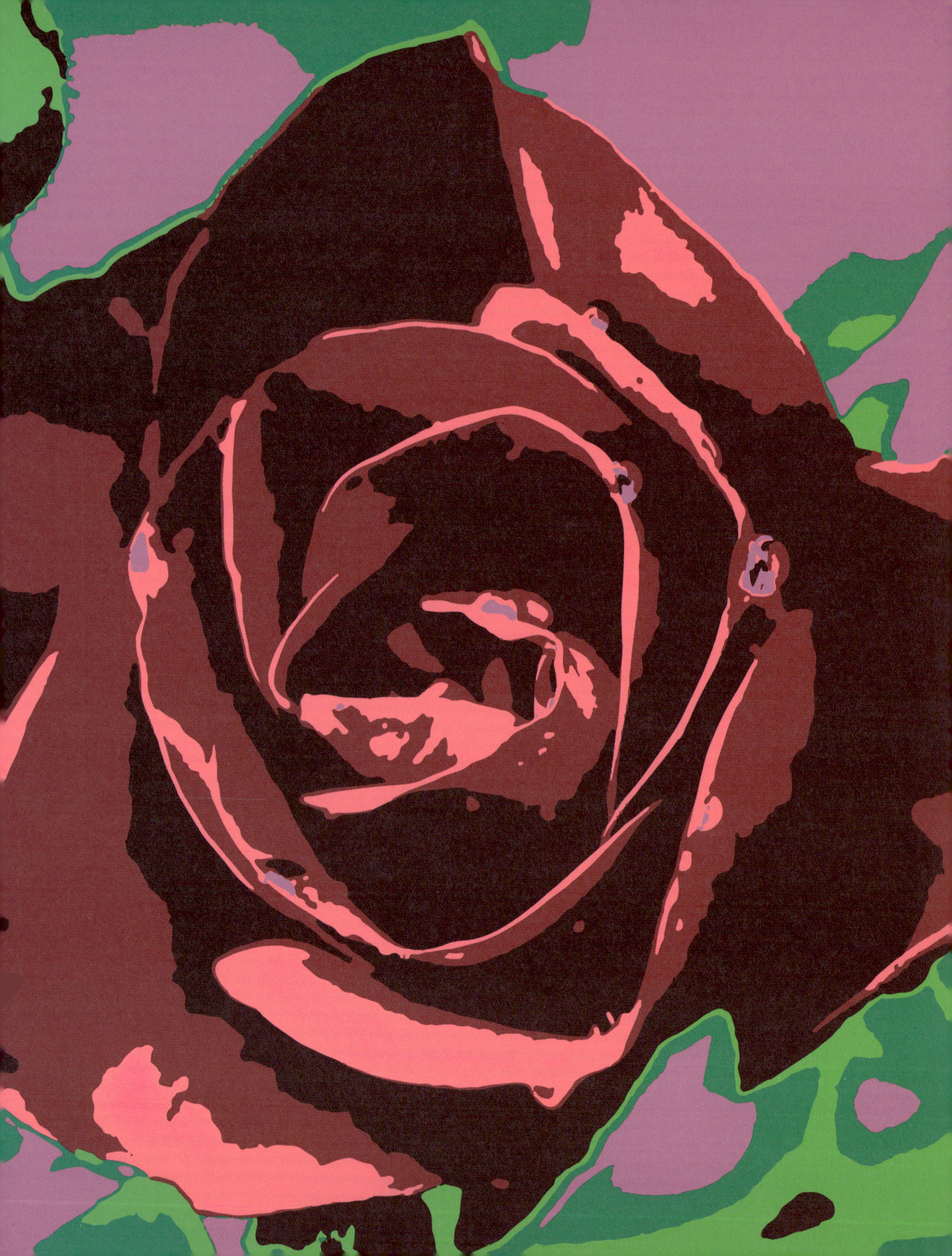

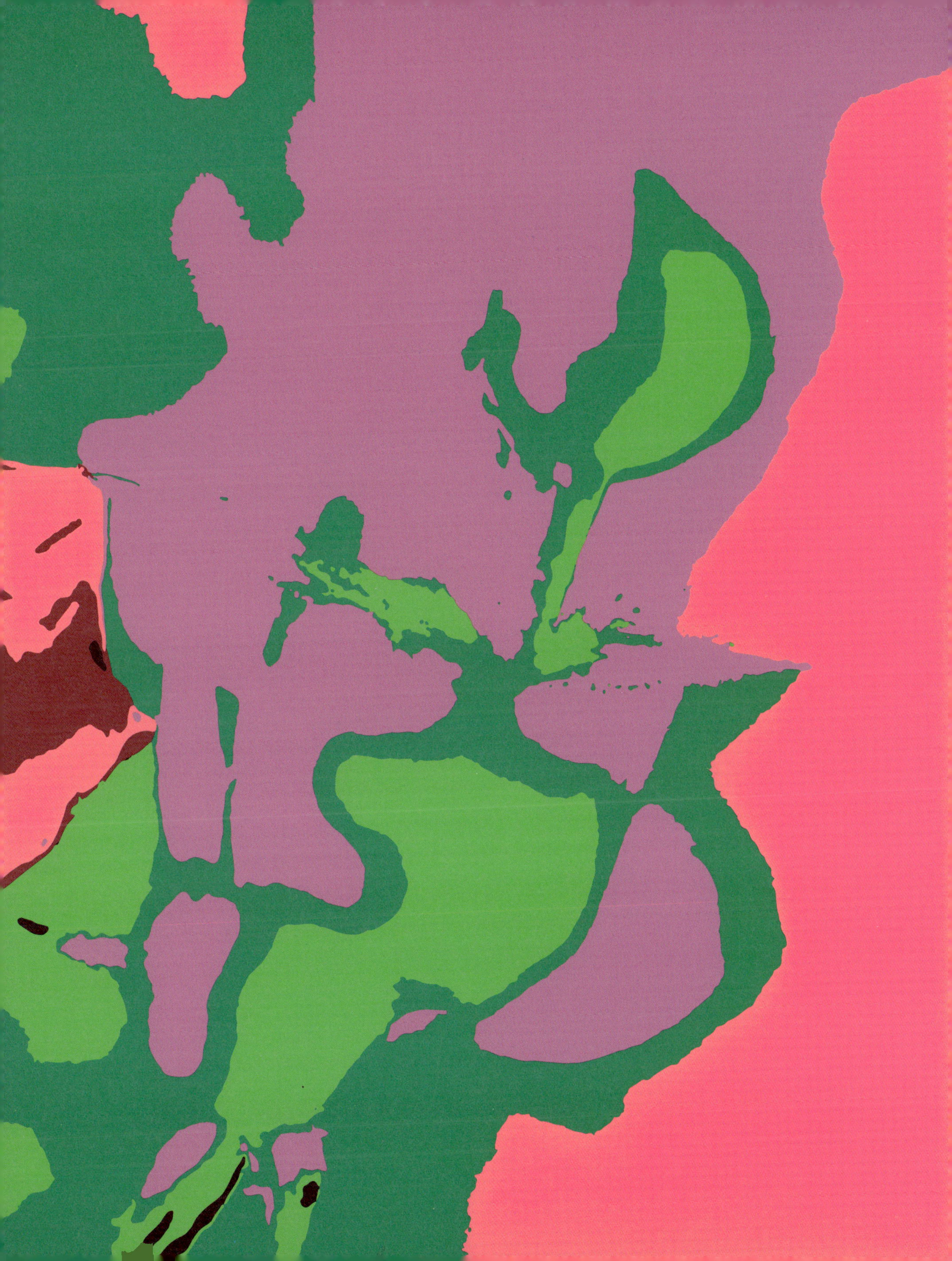

CONTENT
INHALT

CONTENT | INHALT

THE LANGUAGE OF ROSES
DIE SPRACHE DER ROSEN

THE ROSE IN ART
DIE ROSE IN DER KUNST

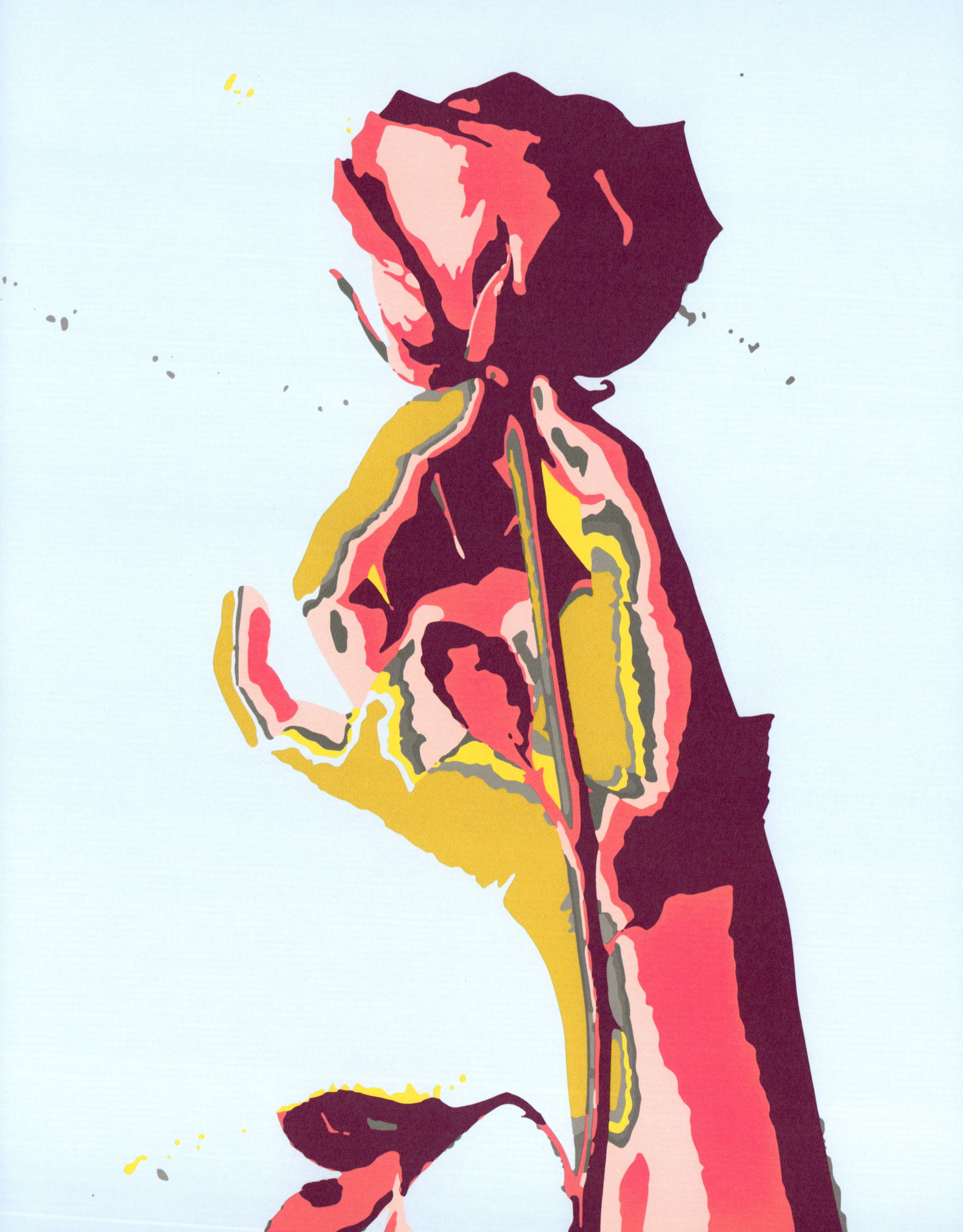

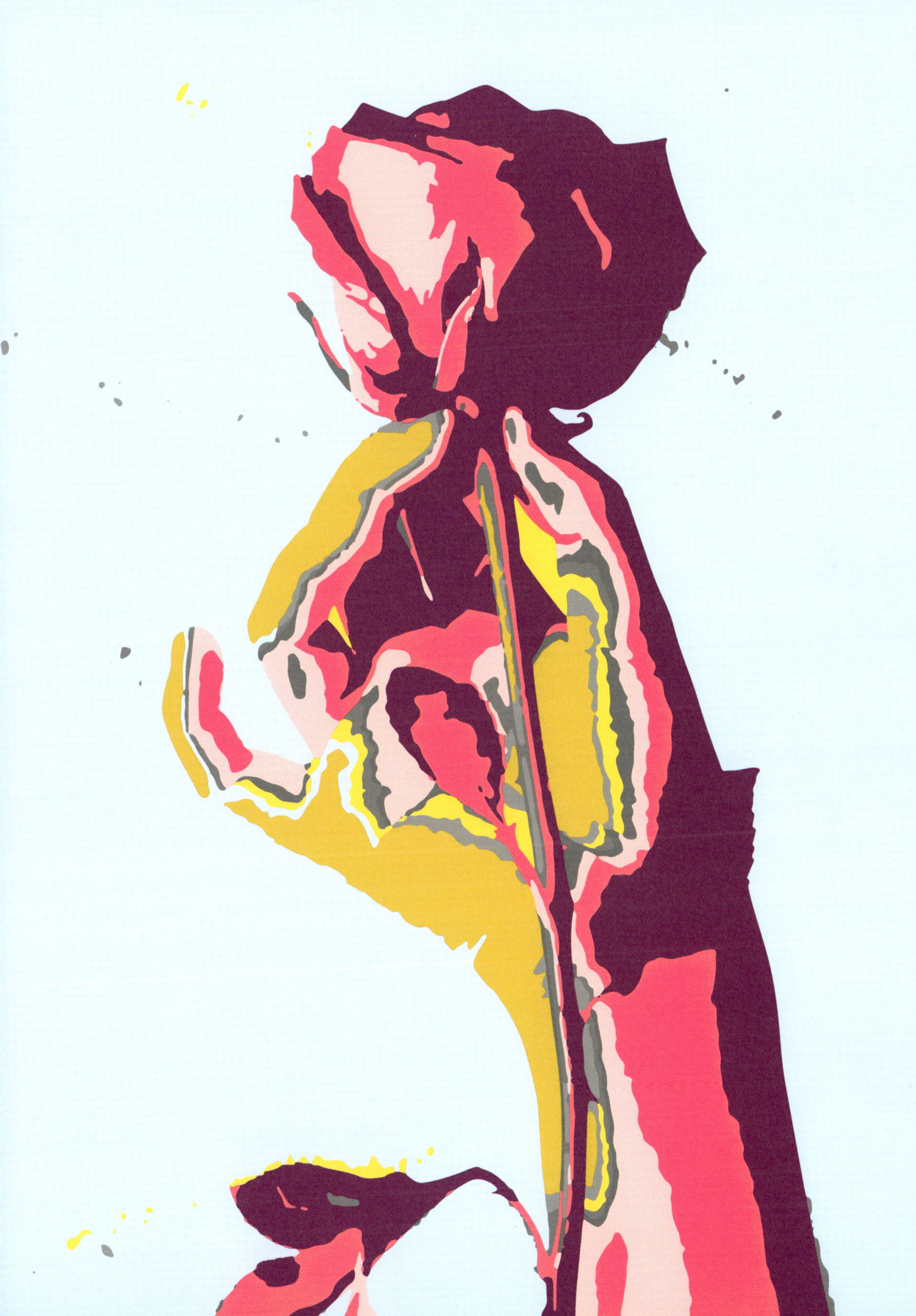

MOMENTS WITH ROSES, MOMENTS OF JOY

My earliest moments with roses lie in childhood. Every corner of my mother's garden was abloom with roses of the most various kinds. Here a 'Queen Elizabeth', an elegant and unrelenting budder, there a delicate moss-rose with its playful pink buds. My very special personal favorite, however, was 'The Squire', an English rose. What I remember about it even more than its dark red, nearly black-satin flowers is its intoxicating fragrance. Even now, I can remember it as vividly as if it was only yesterday that I was whiffing these flowers. Since then, roses have grown on me even more, and I would always encounter them outside the garden, too: in art and literature; in culinary treats; now and then, even as a surprise protagonist in history-making events.

For rose-friends who may wish to learn an inspiring, exciting, or maybe even weird thing or two about the "queen of flowers", I have collected here some of my most favorite impressions of the rose.

May it please you to explore and enjoy these wonderful moments with roses.

Anja Klaffenbach

ROSENMOMENTE – GLÜCKSMOMENTE

Meine ersten Rosenmomente liegen in meiner Kindheit: Im Garten meiner Mutter blühten an allen Ecken die unterschiedlichsten Rosen – hier die elegante, unermüdlich blühende *Queen Elizabeth*, dort eine zarte Moosrose mit verspielten rosafarbenen Knospen. Ganz besonders hatte es mir aber die Englische Rose *The Squire* angetan: Mehr als ihre dunkelroten, ja fast schwarzen samtigen Blüten ist mir ihr betörender Duft noch heute in Erinnerung, so intensiv, als hätte ich erst gestern an einer der Blüten geschnuppert. Seither sind mir Rosen ans Herz gewachsen, und immer wieder begegneten sie mir auch außerhalb des Gartens: in Kunst und Literatur, als kulinarische Genüsse oder als bisweilen unerwartete Protagonisten bei Begebenheiten, die Geschichte schrieben.

Für Rosenfreunde, die Inspirierendes, Spannendes und bisweilen Skurriles um die „Königin der Blumen" entdecken möchten, sind hier einige meiner liebsten Rosenimpressionen versammelt.

Viel Freude beim Erkunden und Genießen
von wunderbaren Rosenmomenten!

Anja Klaffenbach

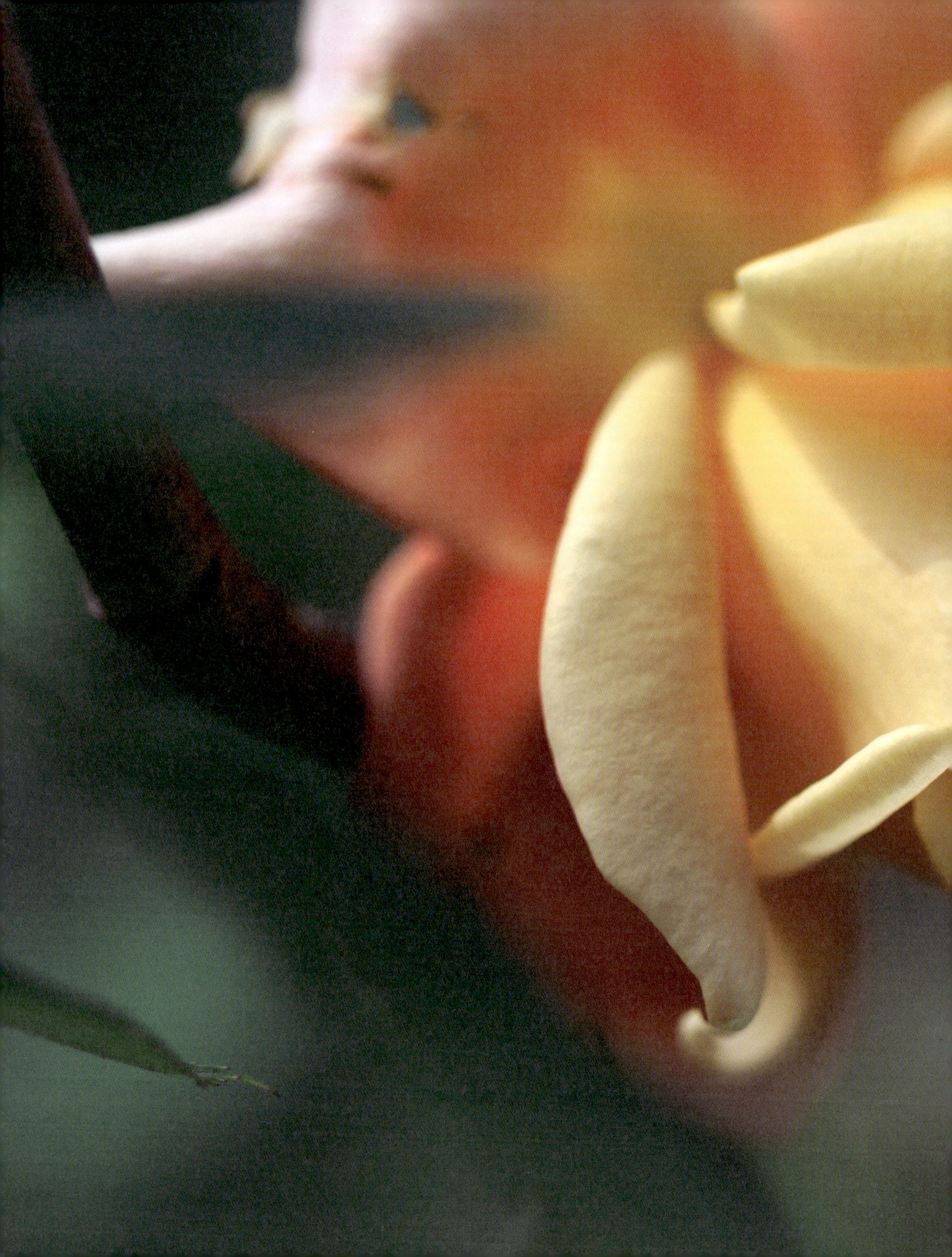

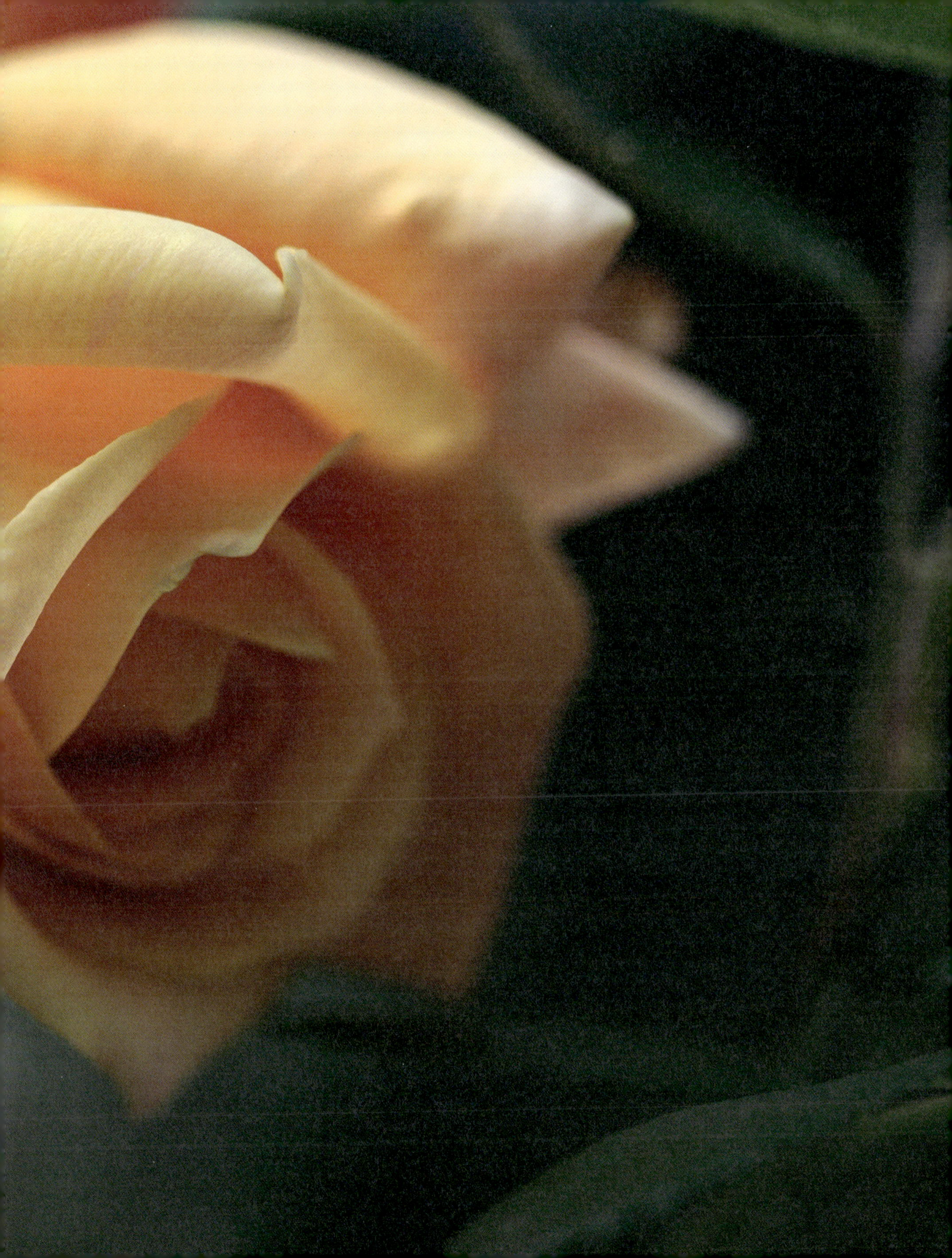

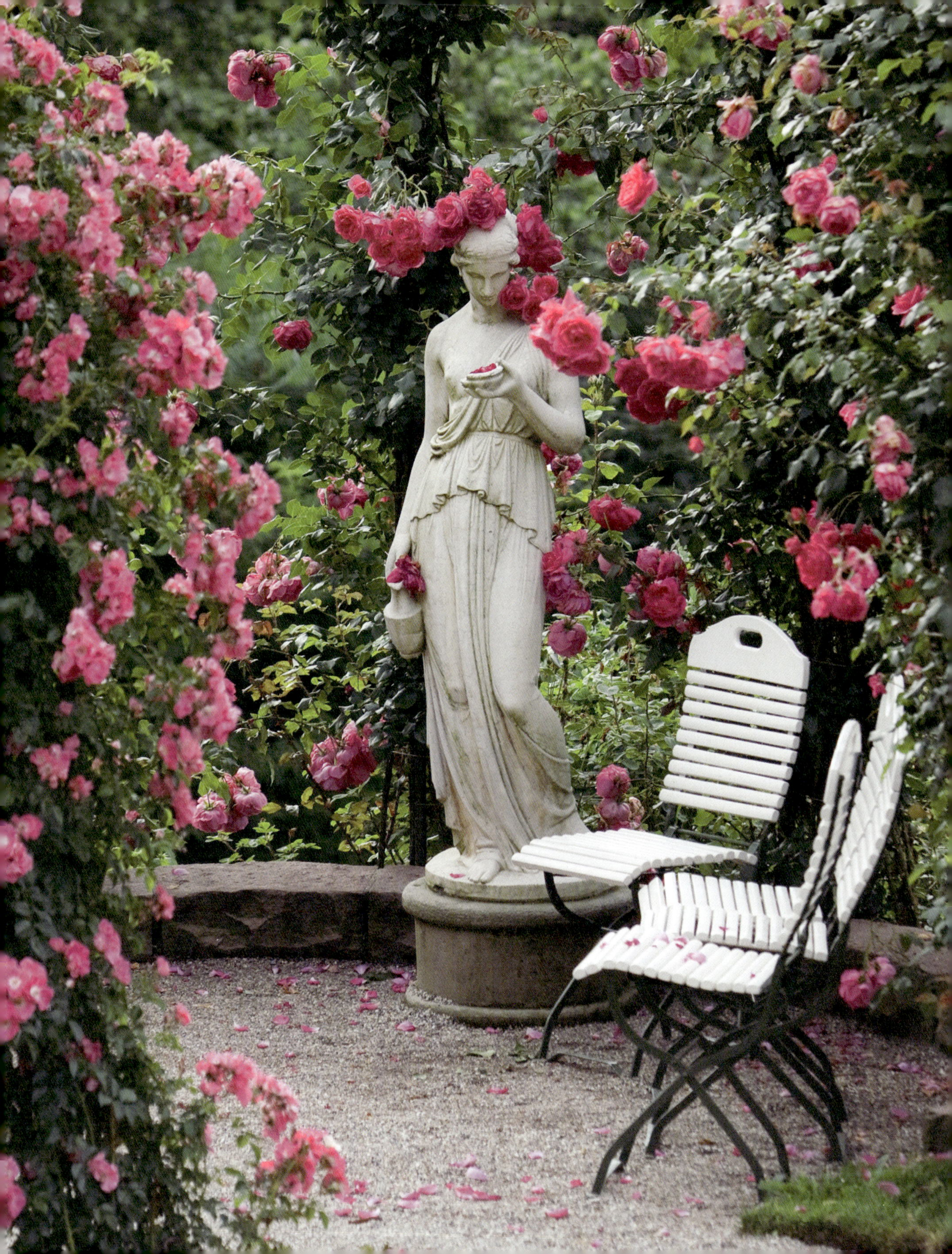

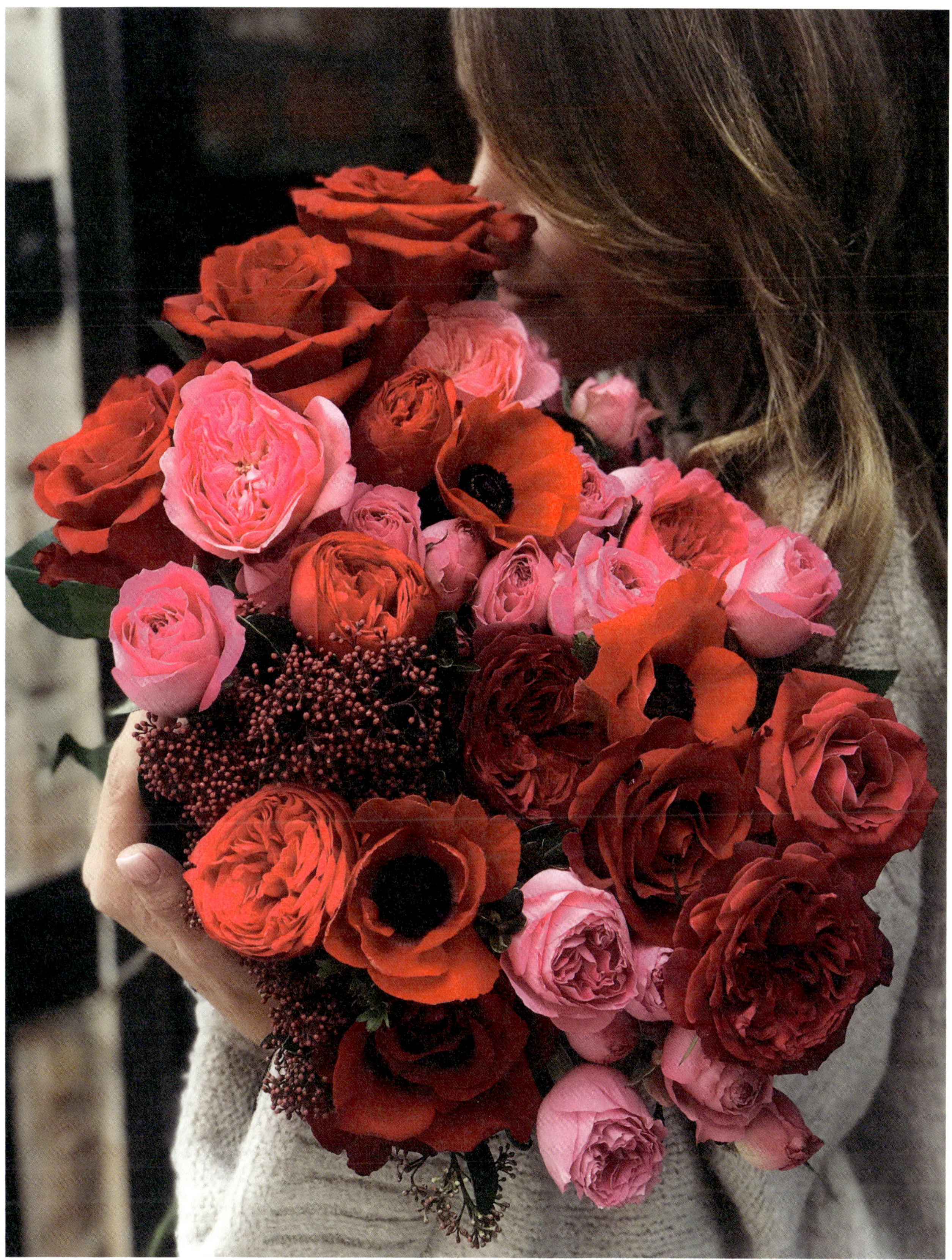

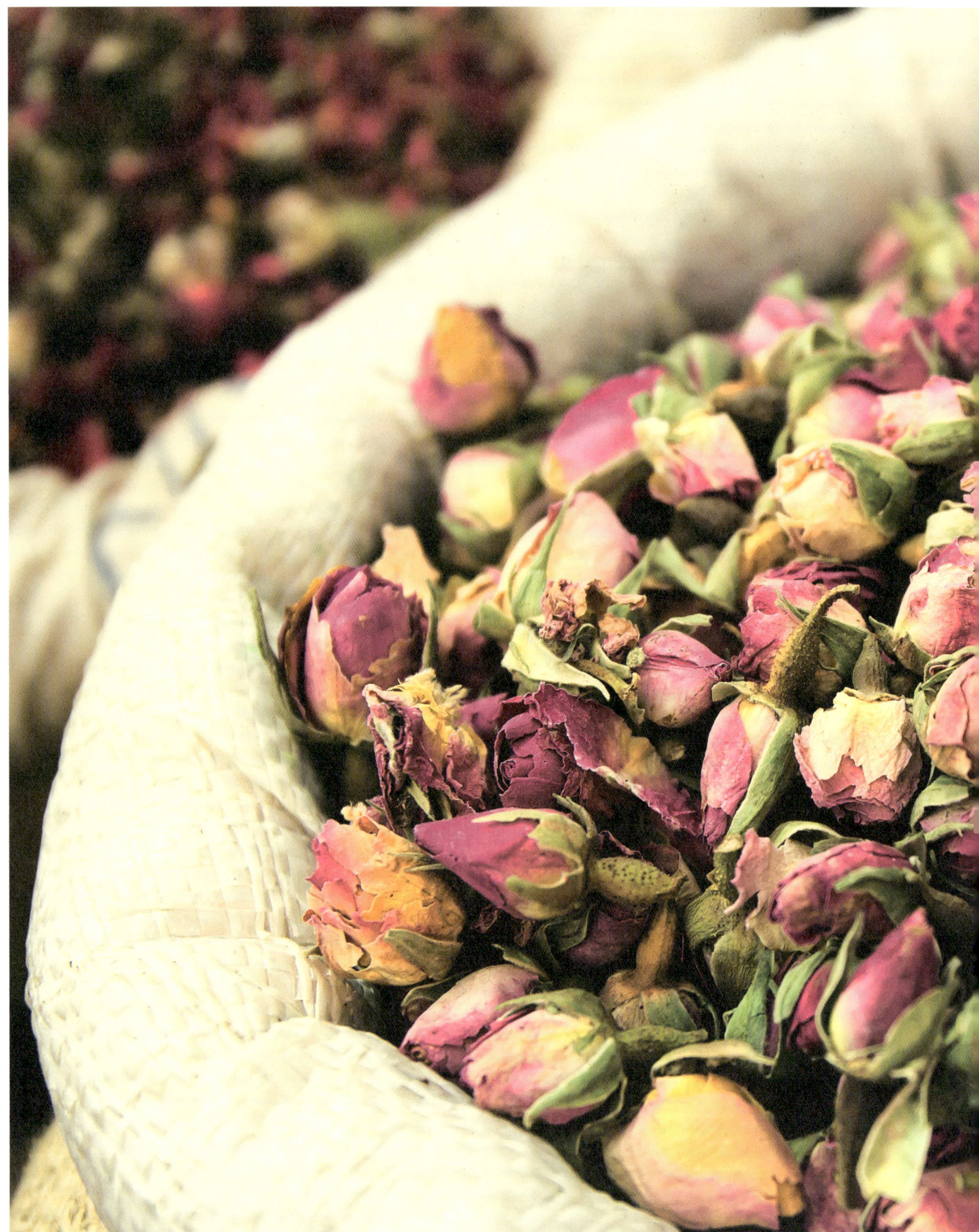

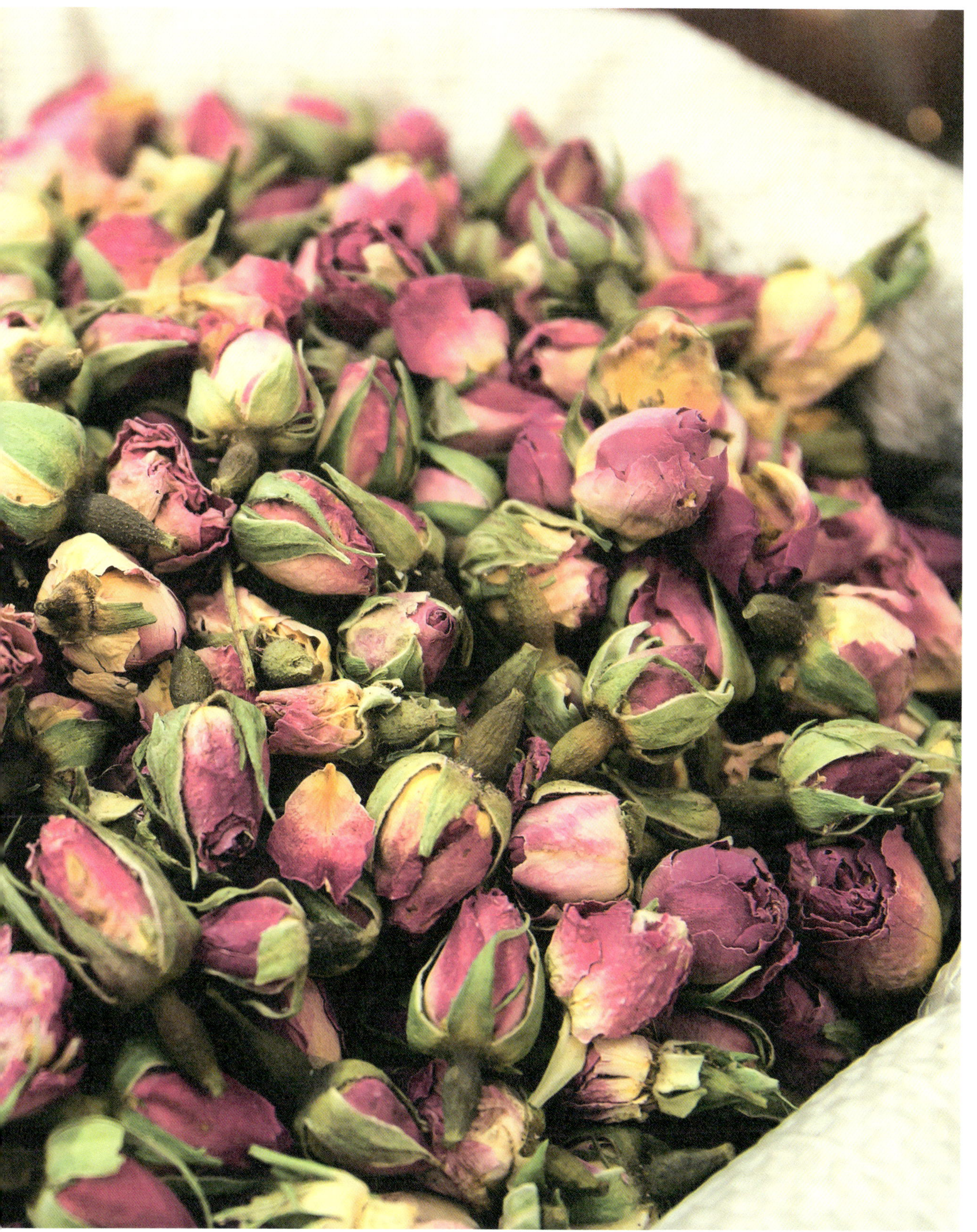

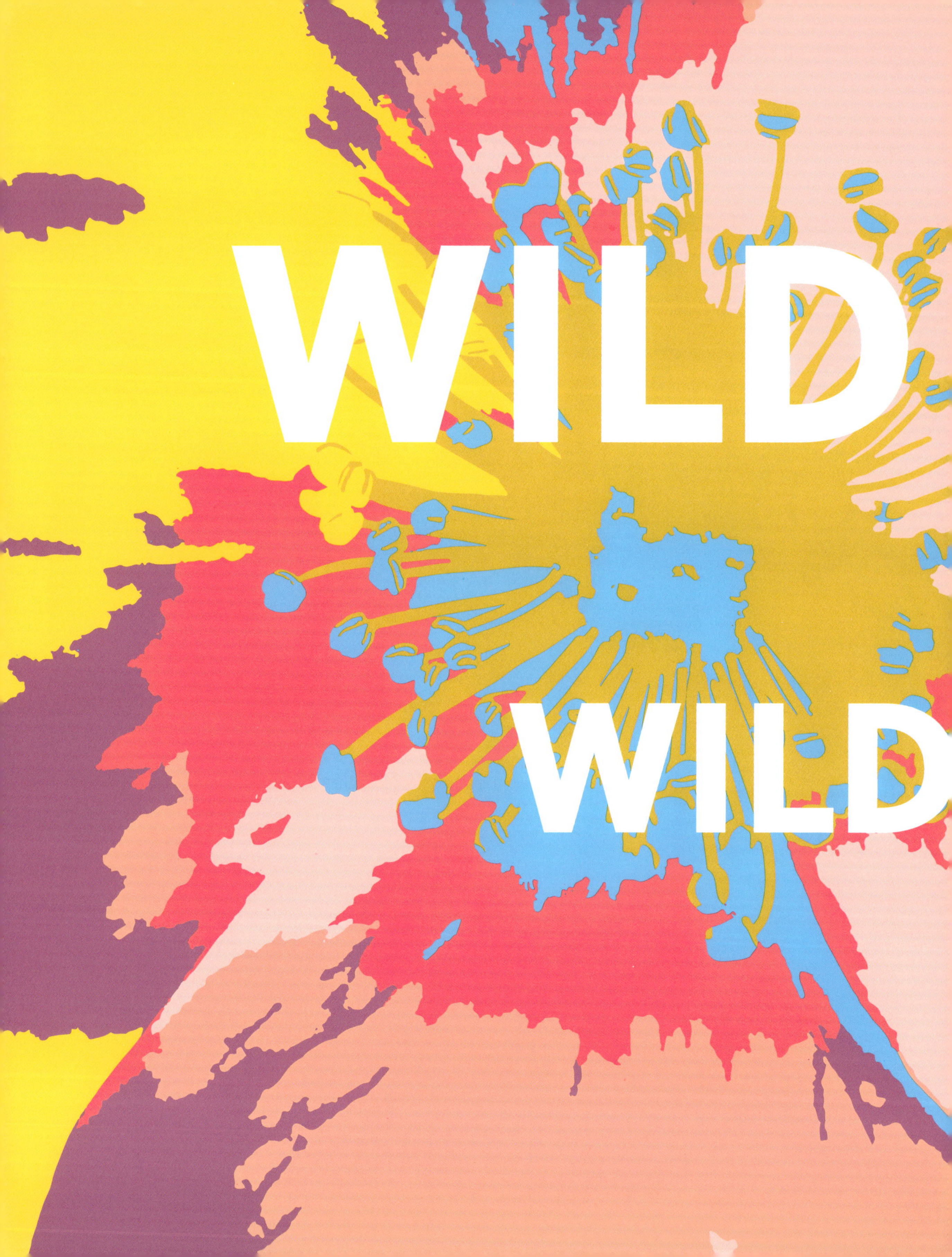
WILD
WILD

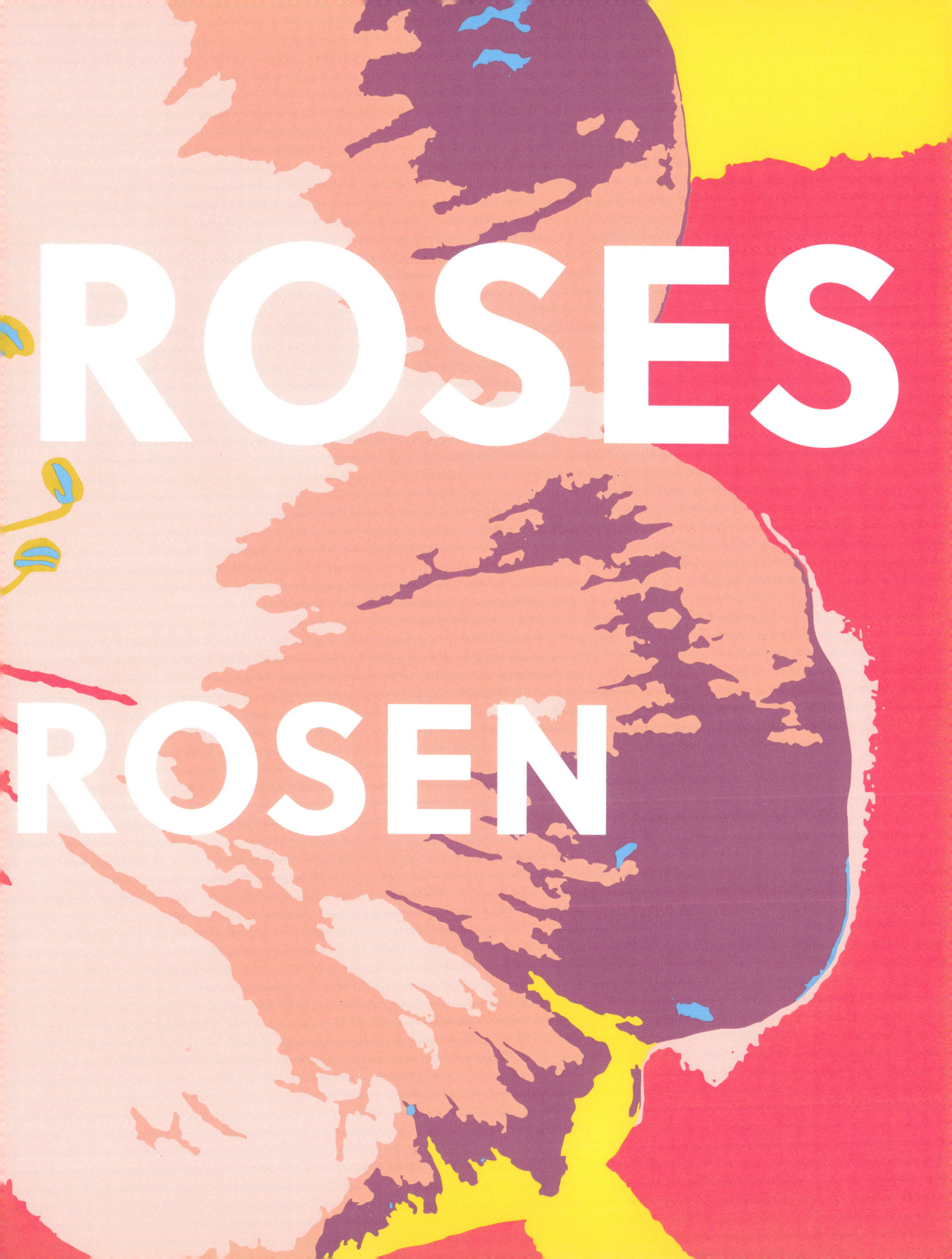
ROSES
ROSEN

Rosa canina

DOG ROSE

INDESTRUCTIBLE

The sweetly fragrant *Rosa canina* is the most common wild rose species in central and southern Europe. In the German-speaking countries it is called dog rose or hedge rose. It blossoms only once, in mid-summer, but with stamina; usually a baby pink, although some variations may be white or a flusher pink shade. With its curved prickles it latches onto other woody plants and thereby attains heights of three to five meters—very tall growth for such a long-lived wild rose. It is no rarity for a single briar to live to be three hundred years old. Charlemagne (p. 128) is known to have treasured *Rosa canina* for its vitamin-rich rose hips and to have decreed that it be cultivated at courts and palaces throughout the realm.

Whether the name "dog rose" comes from being used to heal dog bites or its *dog*-ishness was more of a dig at this common plant's unchecked roadside sprawl, none can say. Probably the most famous *Rosa canina* grows at St. Mary's Cathedral in Hildesheim: The Thousand-year Rose, they call it, is associated with the legend of the cathedral's ninth-century origins. This specimen has truly demonstrated the will to live: Although it burned when the cathedral was bombed in 1945, it was not long before new shoots grew out of its still-intact roots—a modern miracle of roses (p. 148) for the religious and for flower-lovers alike.

NEXT PAGE | A perhaps 1000-year-old rose bush at St. Mary's Cathedral, Hildesheim (right page)

Rosa canina

HUNDSROSE

DIE UNVERWÜSTLICHE

Die lieblich duftende *Rosa canina* ist die in Mittel- und Südeuropa am weitesten verbreitete Wildrosenart und bei uns als Hundsrose oder Heckenrose bekannt. Sie blüht im Hochsommer einmalig, aber ausdauernd, meist in blassem Rosa, in manchen Variationen auch in Weiß oder intensivem Pink. Mit ihren gebogenen Stacheln hakt sie sich gerne in andere Gehölze ein, das macht sie mit drei bis fünf Metern zu einer sehr hoch wachsenden Wildrose, die auch langlebig ist: Ein Strauch kann gut und gerne dreihundert Jahre alt werden! Schon Karl der Große (s. S. 132) schätzte die *Rosa canina* mit ihren vitaminreichen Hagebutten und verfügte ihren Anbau in allen Pfalzen seines Reiches.

Ob der Name Hundsrose darauf zurückgeht, dass sie zur Heilung von Hundebissen verwendet wurde, oder eher das „Hündische" abschätzig für eine gewöhnliche, unbeachtet am Wegesrand vor sich hin wuchernde Pflanze steht – wer weiß? Die wohl berühmteste *Rosa canina* wächst am Mariendom in Hildesheim empor: Den „Tausendjährigen Rosenstock" verbindet man mit der Gründungslegende des Doms im 9. Jahrhundert. Überlebenswillen hat dieses Exemplar wahrlich bewiesen: Als der Dom im Jahr 1945 zerbombt wurde und der Rosenstock verbrannte, kämpften sich schon kurz darauf neue Triebe aus der noch erhaltenen Wurzel nach oben – für Gläubige und Blumenfreunde ein modernes Rosenwunder (s. S. 150)!

NÄCHSTE SEITE | Womöglich 1000-jähriger Rosenbusch am Mariendom in Hildesheim (rechts)

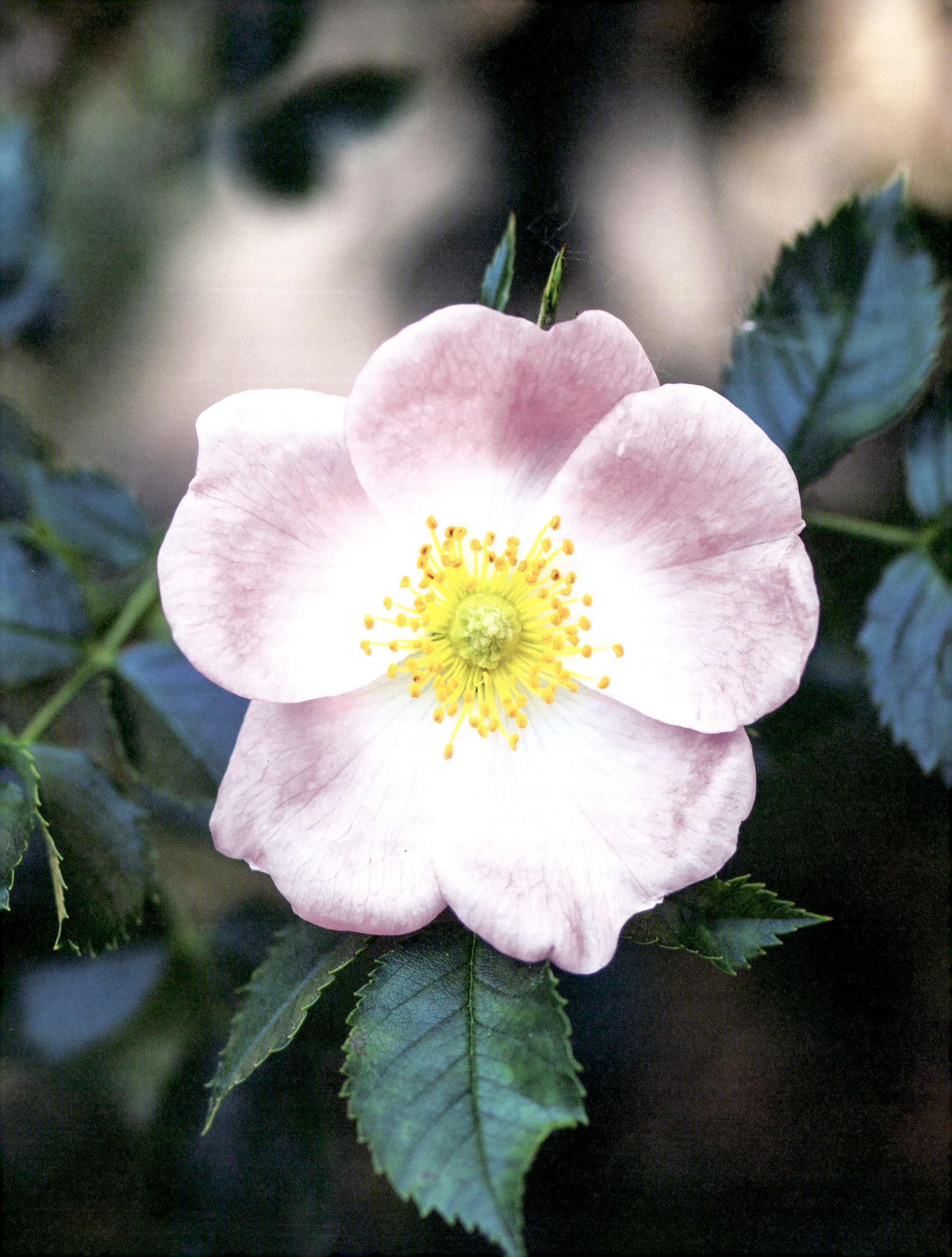

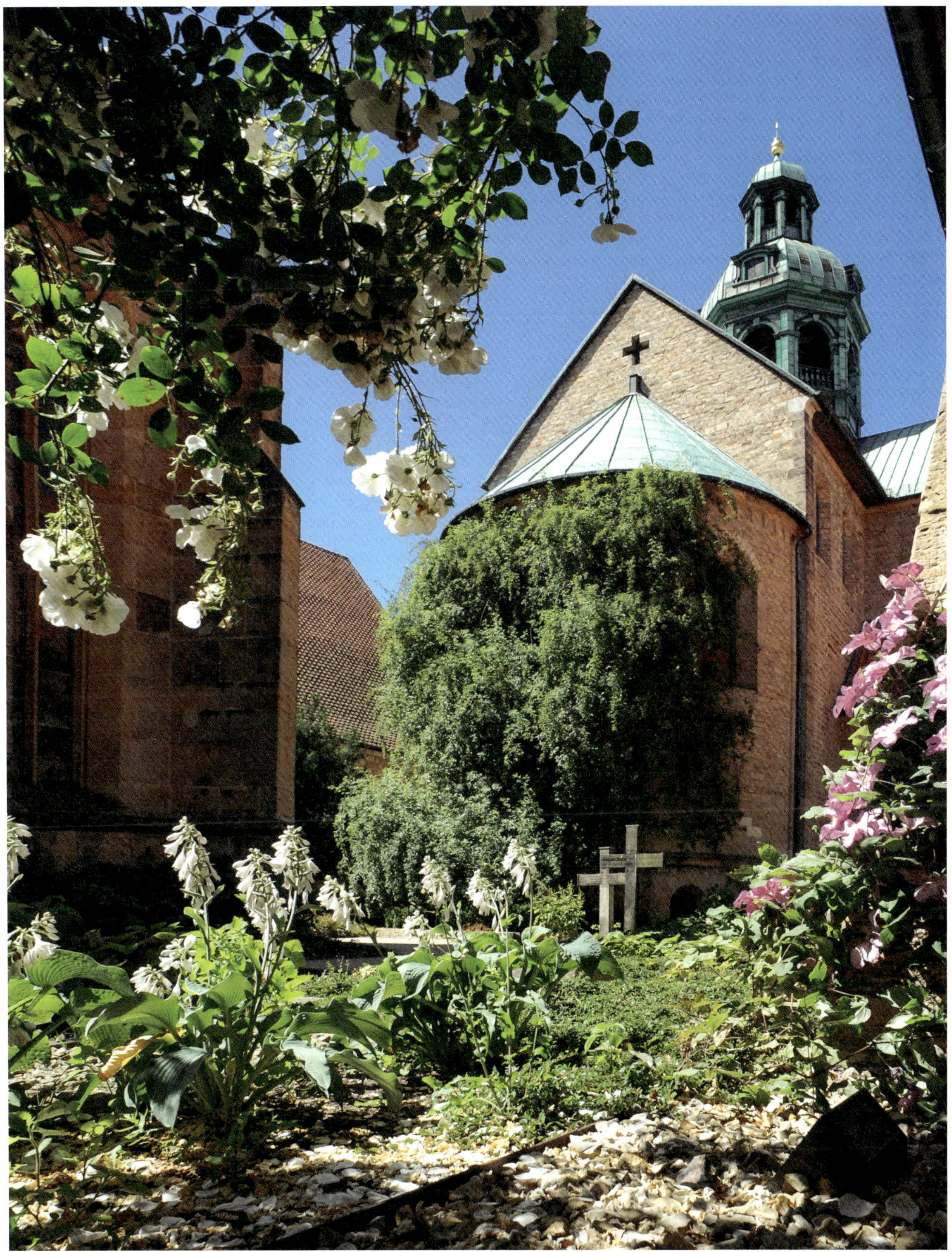

Rosa pimpinellifolia

SCOTCH OR BURNET ROSE

NORDIC BY NATURE

One of the oldest native species of wild rose, *Rosa pimpinellifolia* is extremely hardy. The Scotch (or burnet) rose loves the barren, sandy soils along the North Sea coasts and does perfectly well with salt and wind, which is why many people also know it as the dune rose. This coastal beauty has a rather prickly presentation, and so another of its monikers is *Rosa spinosissima* (Latin for "thorny, prickly"). After it blossoms, its glinting black hips make this rose a stand-out fall decorative.

BIBERNELLROSE

NORDISCH BY NATURE

Äußerst genügsam ist die *Rosa pimpinellifolia*, eine der ältesten heimischen Wildrosenarten: Die Bibernellrose liebt die kargen sandigen Böden der Nordseeküste und kommt mit Wind und Salz bestens zurecht – im Volksmund ist sie daher auch als Dünenrose bekannt. Die Küstenschönheit gibt sich zudem überaus stachelig, weshalb sie auch als *Rosa spinosissima* (lat. für „dornig, stachelig") bezeichnet wird. Und nach der Blüte machen ihre schwarz glänzenden Hagebutten die Bibernellrose zum aparten Schmuckstück im Herbst.

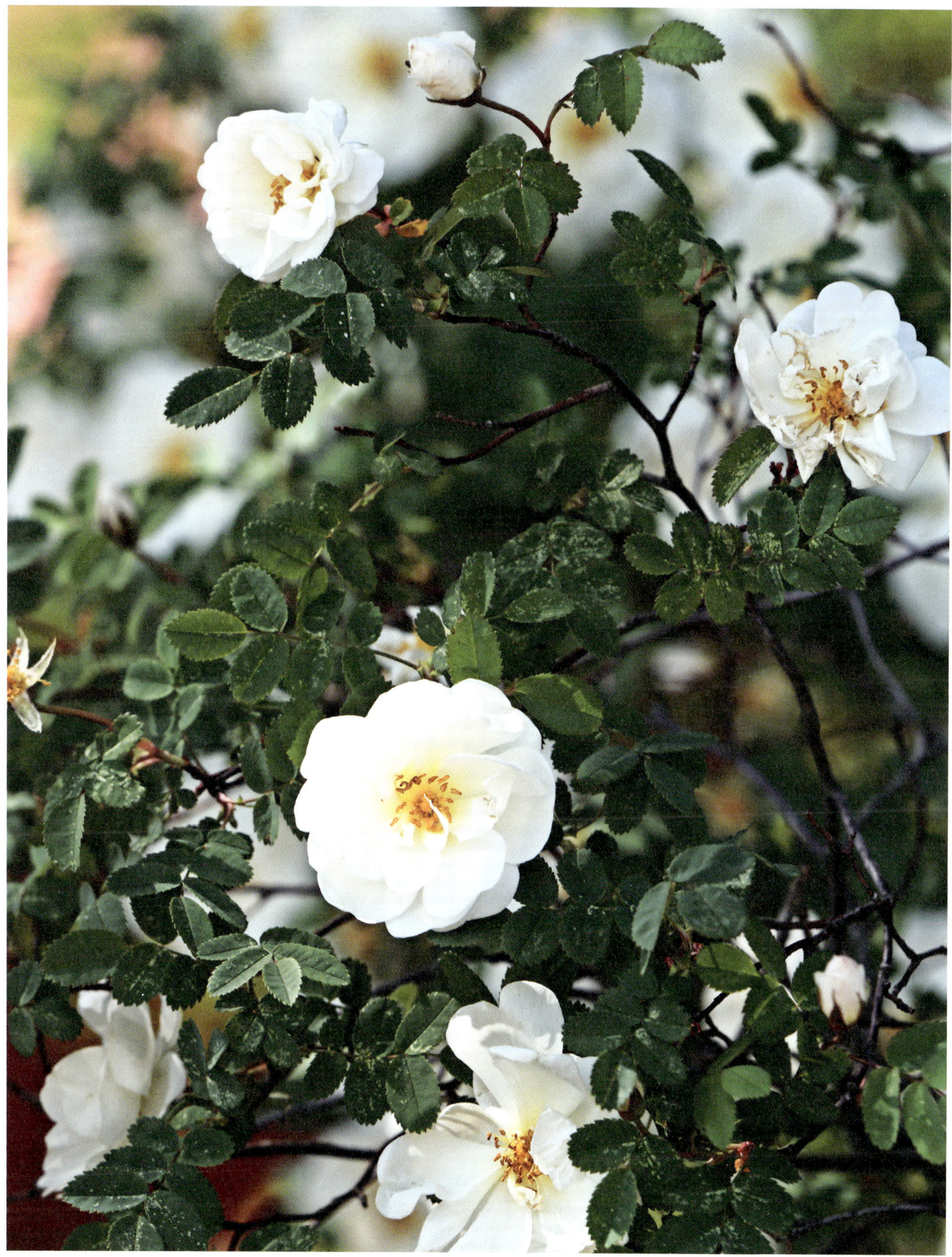

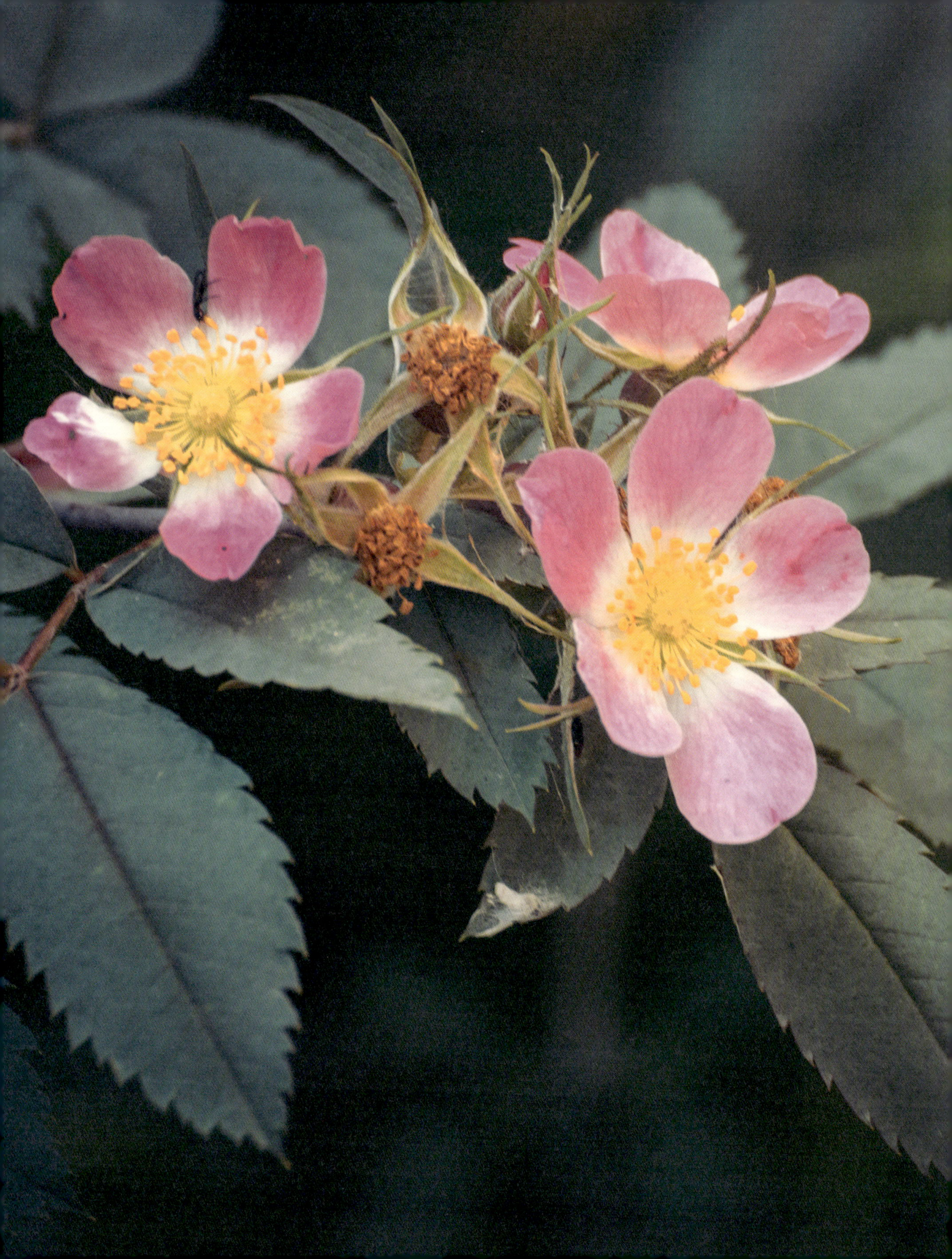

Rosa glauca

RED-LEAVED ROSE

DELICATE FLOWERS, HANDSOME FOLIAGE

Rosa glauca is a purple-flowering wild rose that prefers nutrient-poor soils in mountainous regions of central Europe and can withstand heavy frosts. Despite that, it is regarded as threatened, especially in southern Germany. The leaves and stems are very attractive, coated in just a breath of blueish green, which is why this rose is popularly called the pike rose. The new shoots of this rose have a reddish tint, and hence its Latin synonym, *Rosa rubrifolia*.

HECHTROSE

ZARTE BLÜTEN, SCHMUCKES LAUB

Die *Rosa glauca* ist eine purpurfarben blühende Wildrose, die vornehmlich in süd- und mitteleuropäischen Gebirgsregionen auf kargen Böden wächst und auch heftigen Minusgraden trotzt – dennoch gilt sie vor allem in Süddeutschland als gefährdet. Sehr reizvoll sind die zart in Hechtblau überhauchten Blätter und Stängel, die zum volkstümlichen Namen Hechtrose führten. Da die jungen Triebe rötlich gefärbt sind, ist sie auch als *Rosa rubrifolia*, rotblättrige Rose, bekannt.

Rosa rugosa

BEACH ROSE

THE INVASIVE BEAUTY

Its corrugated leaves, which look like the foliage of the potato, is the reason *Rosa rugosa* is commonly called the *Kartoffelrose* in German. Its open flowers are highly decorative in hot pink, baby pink, or white. Also sought-after are their large, luminous red fruits, often worked into a rose-hip paste, syrup, or butter. In Germany, *Rosa rugosa* is also often called the "Sylt rose", after the island of Sylt; but this is not its native habitat. This rose's other common names, the Japanese or Kamchatka rose, reveal its actual origin in East Asia.

The Swedish biologist Carl Peter Thunberg brought it to Europe in 1796. No amount of salt or wind nor extreme temperatures are a match for *Rosa rugosa*. It loves sandy soils, where its root system races out.

It is both a blessing and a curse: In the northern German coastal areas, beach roses were planted intentionally, to stabilize bands of sand dunes, and they did well enough there to become a favored plant in hedges and windbreaks from the East Frisian islands up through Denmark. Because of its near-uncontrolled spread, however, it has squeezed out native coastal plants such as sandalwood, sea holly, and burnet rose (p. 24). As a result, to prevent their unchecked expansion, Denmark has banned the sale and transfer of these plants.

NEXT PAGE | The good with the bad: *Rosa rugosa* root systems stabilize sand dunes but crowd out other plants.

Rosa rugosa

KARTOFFELROSE

DIE INVASIVE SCHÖNHEIT

Durch ihre gerunzelten Blätter, die an Kartoffellaub erinnern, kam die *Rosa rugosa* zu einem ihrer Volksnamen: Kartoffelrose. Sehr dekorativ sind auch ihre offenen Blüten in Pink, Rosa oder Weiß. Begehrt sind zudem ihre großen, leuchtend roten Früchte, die gerne zu Hagebuttenmark verarbeitet werden. Die *Rosa rugosa* ist auch als Sylter Rose bekannt, doch heimisch war sie dort ursprünglich nicht. Weitere Beinamen dieser Wildrose, Japan-Rose und Kamtschatka-Rose, verraten die eigentliche Herkunft aus Ostasien.

Nach Europa kam sie um 1796 durch den schwedischen Biologen Carl Peter Thunberg. Der *Rosa rugosa* können weder Salz, Wind noch extreme Temperaturen etwas anhaben. Sie liebt sandige Böden, die sie rasend schnell mit Ausläufern durchwurzelt.

Darin liegen nun Fluch und Segen zugleich: In den nordischen Küstenregionen wurde die Kartoffelrose gezielt zur Befestigung von Dünenstreifen gepflanzt und mauserte sich von den Ostfriesischen Inseln bis nach Dänemark zur beliebten windschützenden Heckenpflanze. Doch mittlerweile verdrängt sie durch ihre fast unkontrollierbare Ausbreitung viele heimische Küstenpflanzen wie Sanddorn, Stranddistel und Bibernellrose (s. S. 24). Daher hat Dänemark den Verkauf und die Weitergabe dieser Pflanzen inzwischen untersagt, um der ausufernden Verbreitung Einhalt zu gebieten.

NÄCHSTE SEITE | Wohl und Wehe: Die *Rosa rugosa* hält mit ihren Wurzeln Dünenstreifen fest, sie verdrängt aber auch andere Pflanzen.

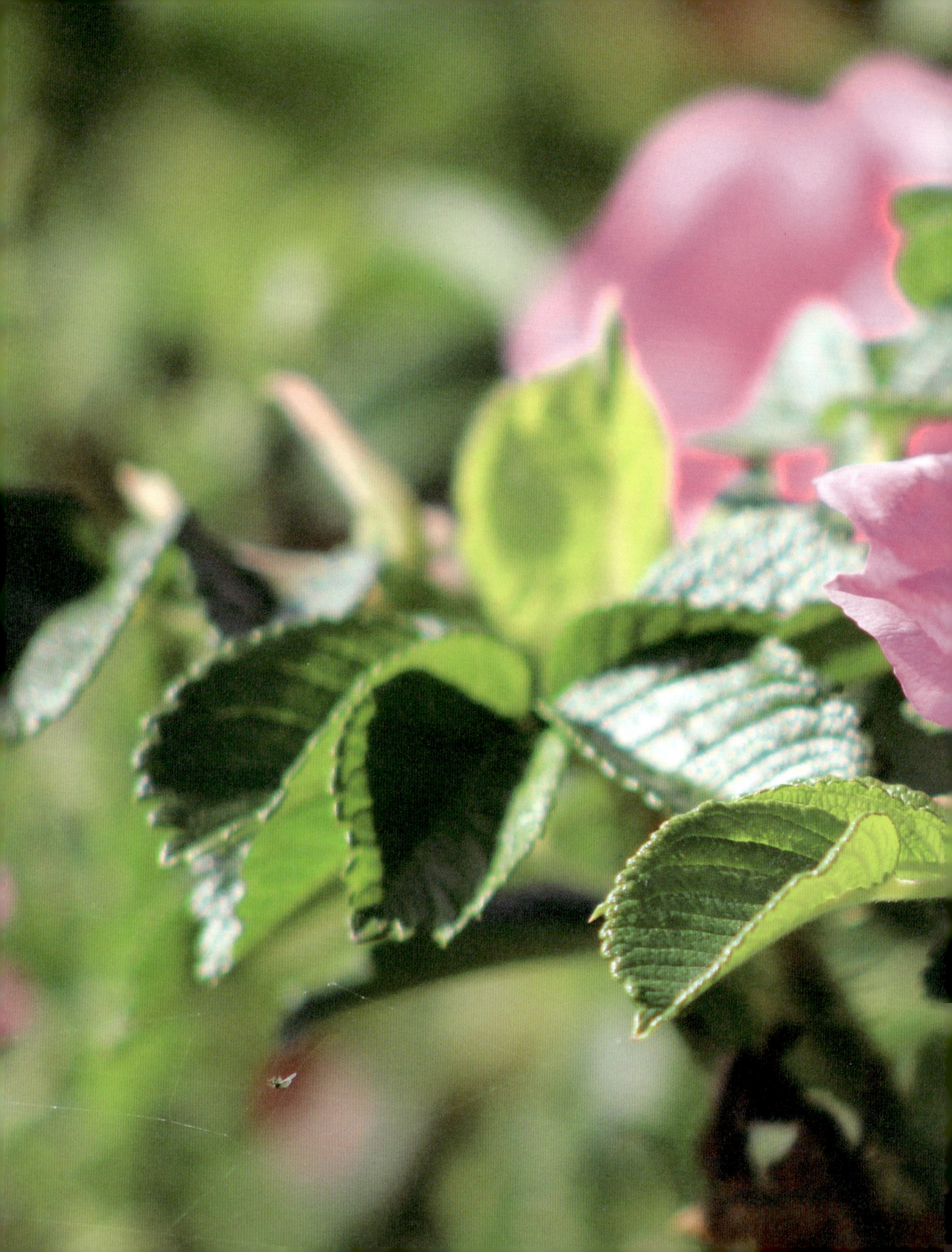

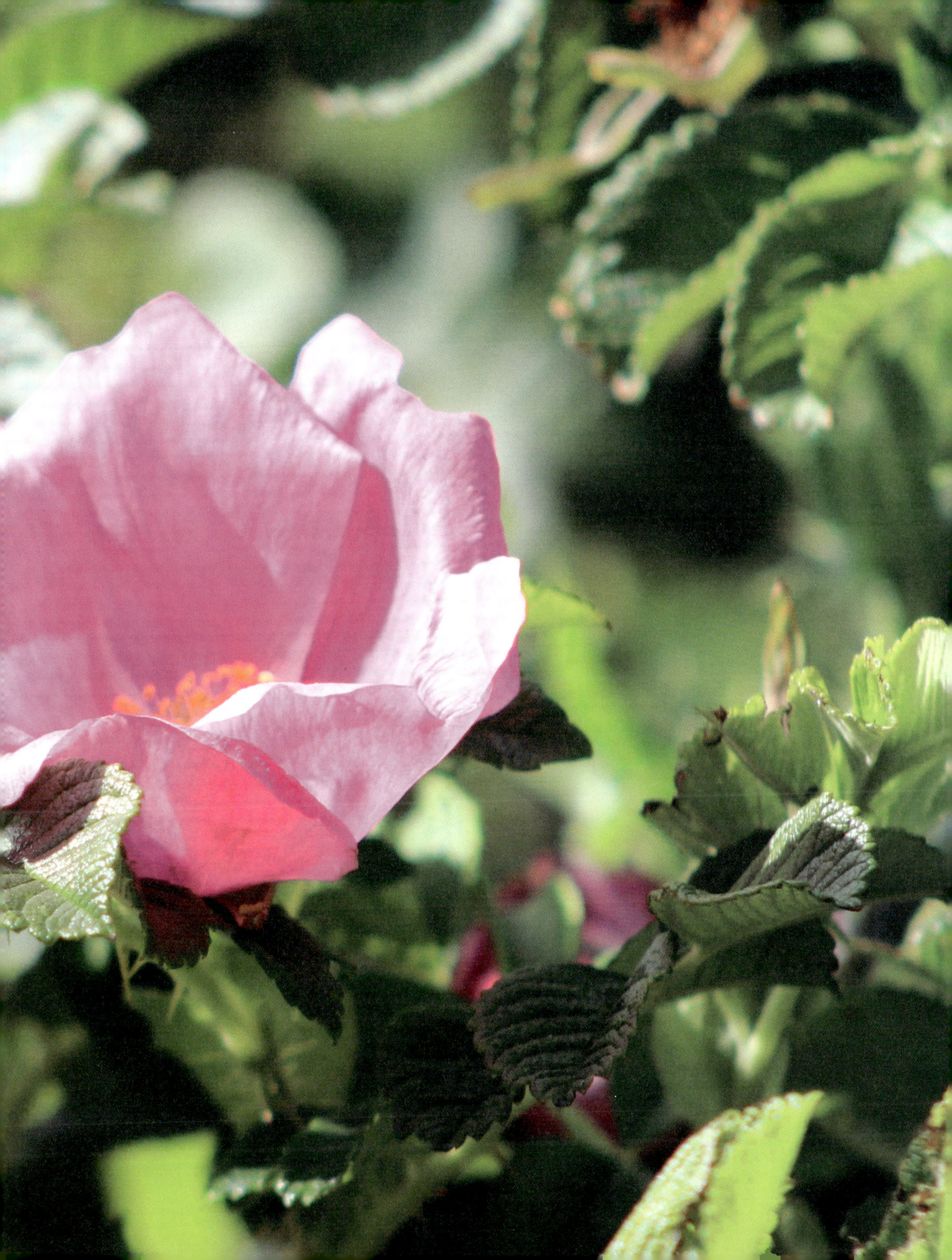

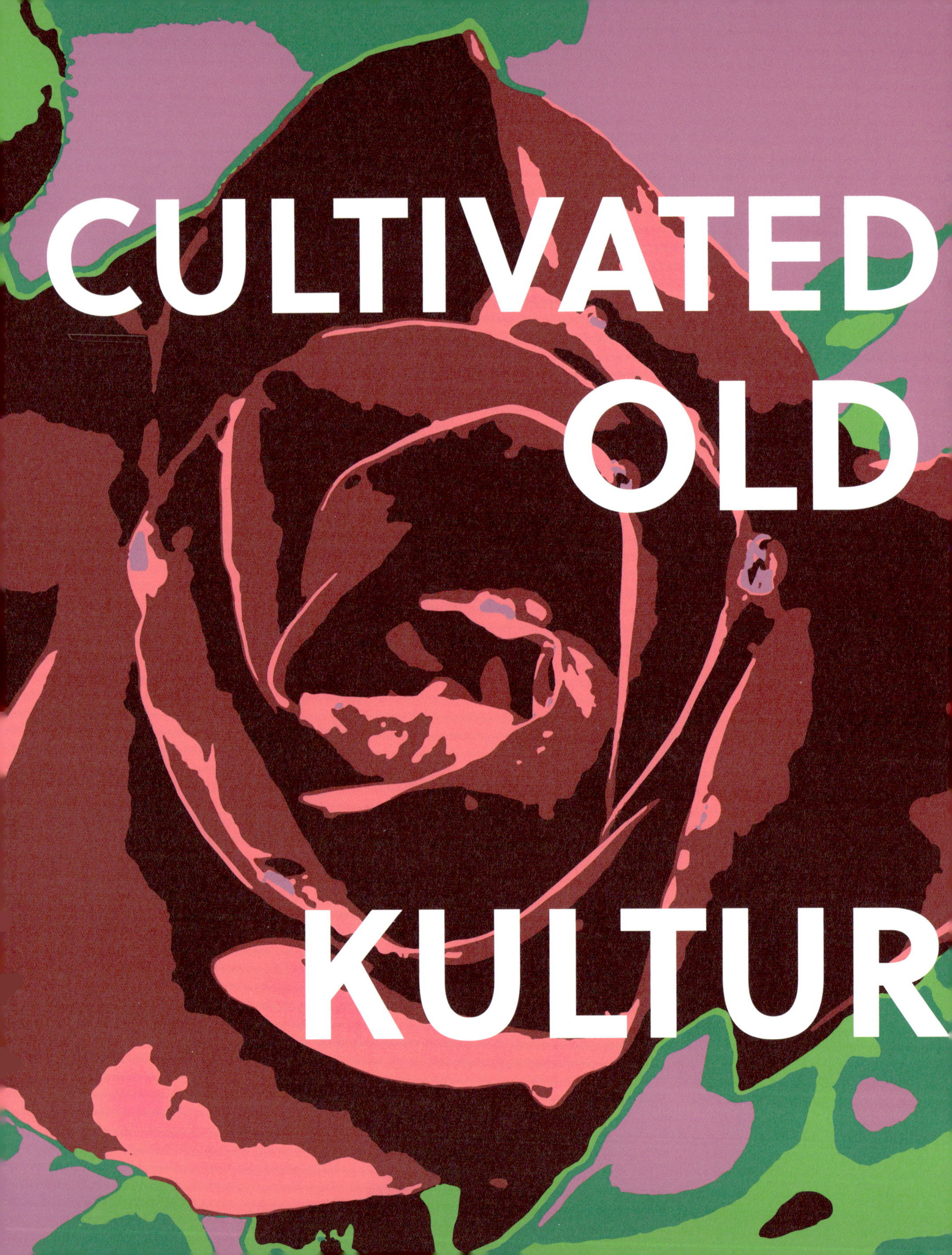

CULTIVATED
OLD
KULTUR

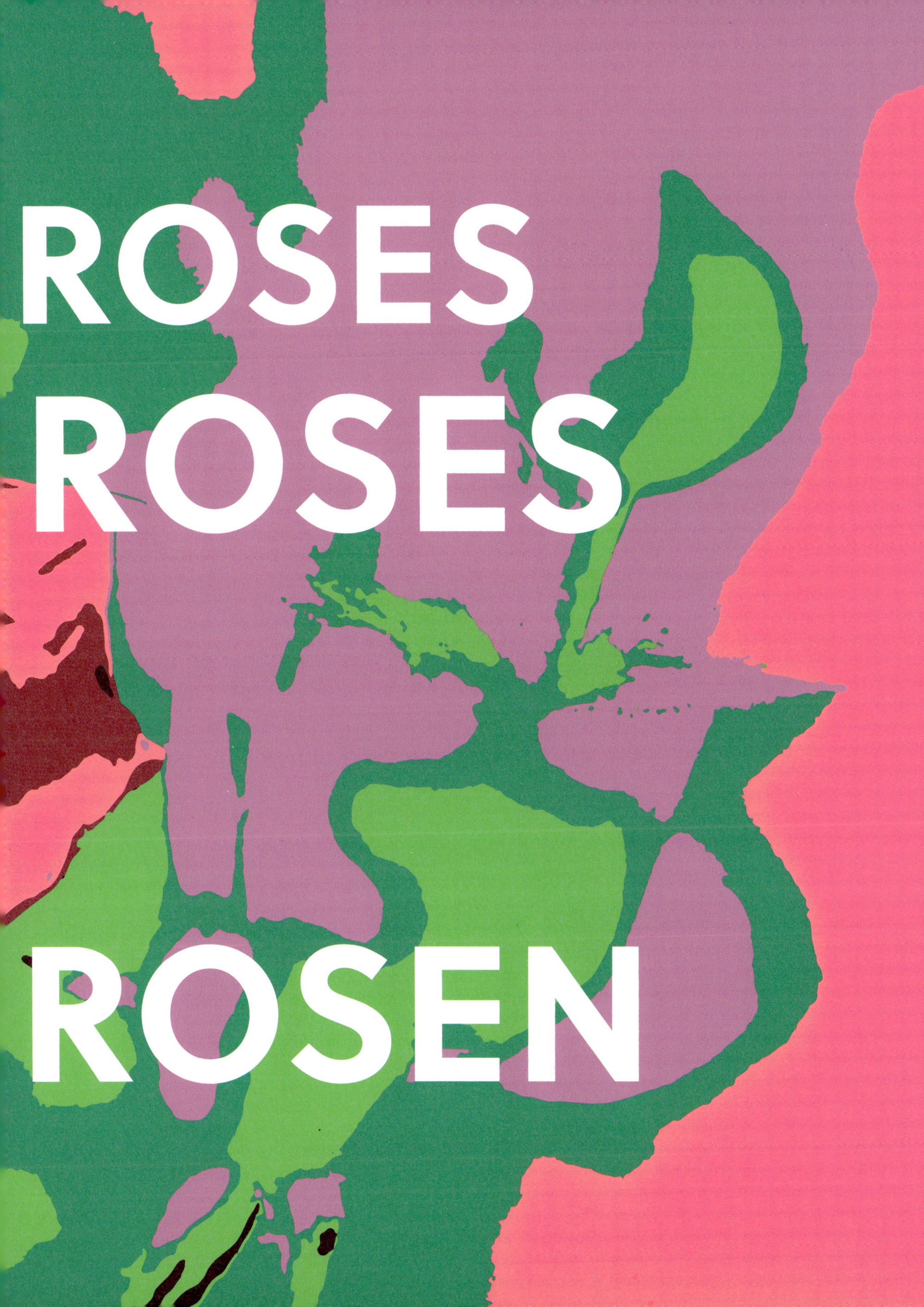
ROSES
ROSES
ROSEN

Rosa gallica

GALLIC ROSE

THE HEALING ROSE

Vivid cherry red, bright purple, saturated pink: *Rosa gallica* likes it strong, the same goes for its intense fragrance. From *Rosa gallica* , rose water, rose oil, and rose vinegar were made for medicinal applications, even in ancient times. The species acquired common names such as the vinegar or apothecary rose. Ruling elites loved *Rosa gallica*, too. It is common in the Mediterranean region; Cleopatra is said to have bathed in Gallic rose flowers; the royal house of Lancaster in England chose it as its emblem.

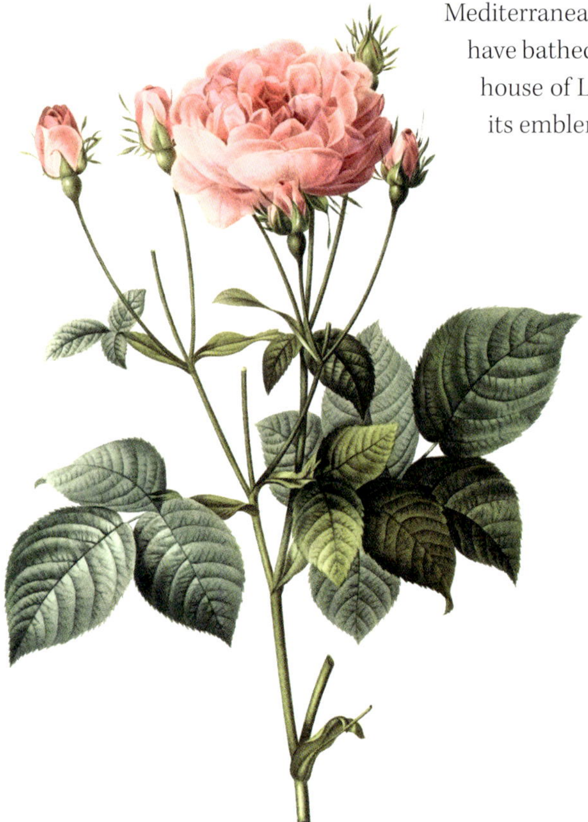

Rosa gallica

ESSIG-ROSE

DIE HEILENDE ROSE

Kräftiges Kirschrot, leuchtendes Purpur und sattes Pink – die *Rosa gallica* liebt es kraftvoll, und das gilt gleichermaßen für ihren intensiven Duft. Schon im Altertum wurden Rosenwasser, Rosenöl und Rosenessig für medizinische Anwendungen aus ihr hergestellt – daher auch die Trivialnamen Essig-Rose oder Apotheker-Rose. Selbst Herrscher schätzten die im Mittelmeerraum verbreitete *Rosa gallica*: Kleopatra soll in ihren Blütenblättern gebadet haben, und in England erkor man die rote Rose zum Emblem des Königshauses Lancaster.

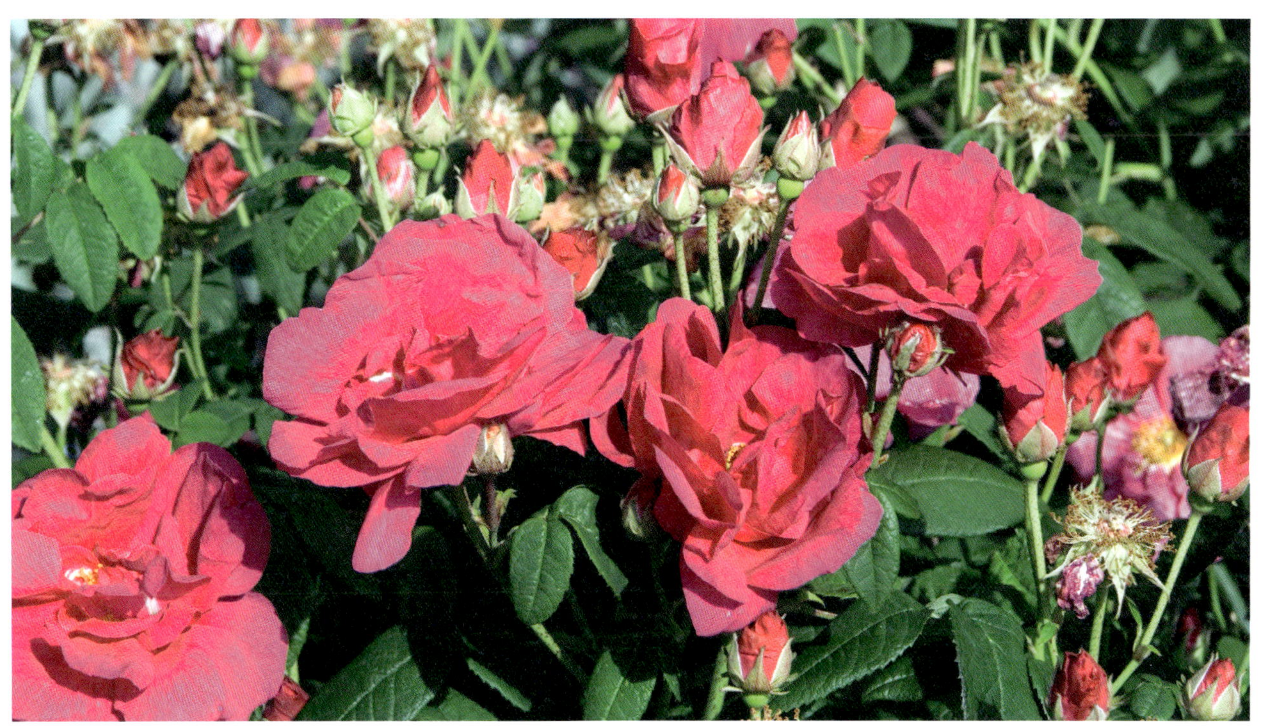

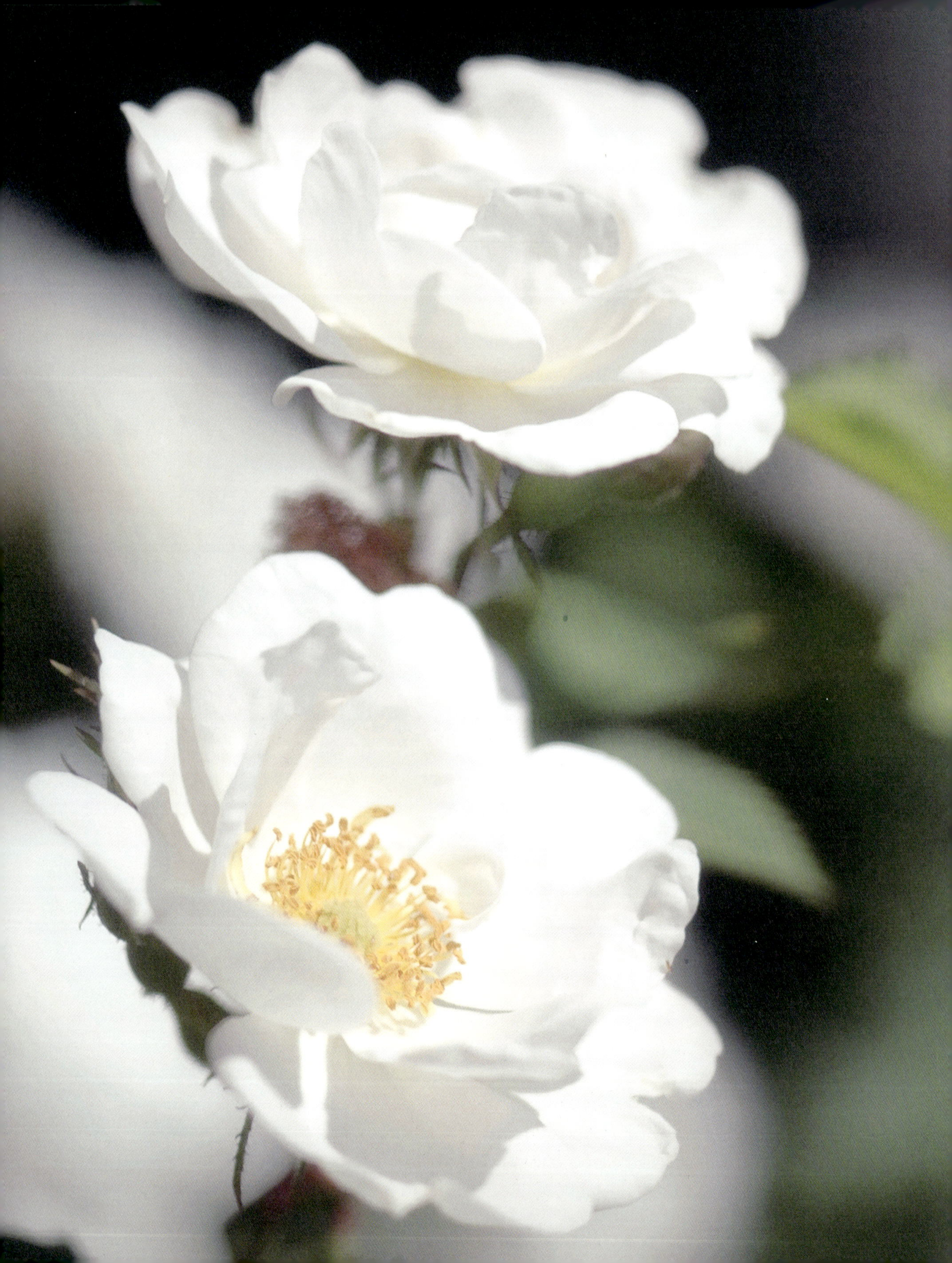

Rosa × alba

WHITE ROSE

FLOWER OF KINGS AND PEASANTS

White flowers with a delicate fragrance; a bushy habit; long stems that may lack prickles: The white rose was already popular in ancient Greece and Rome. Readily cultivated in cottage gardens, *Rosa alba* became known to German speakers as the white "peasant's" rose. And yet the delicate fragrance and noble flower of *Rosa alba* also appealed to a royal sensibility. In England, during the Wars of the Roses, it went down in history as the royal heraldic White Rose of York (p. 112).

WEISSE ROSE

DIE BLUME VON KÖNIGEN UND BAUERN

Zart duftende Blüten in Weiß und ein buschiger Wuchs mit langen, teils stachellosen Stielen: Schon bei Griechen und Römern war die Weiße Rose beliebt. Da sie gerne in Bauerngärten kultiviert wurde, ist die *Rosa alba* auch als Weiße Bauernrose bekannt. Doch selbst königlichen Ansprüchen genügte die zart duftende *Rosa alba*: Als „weiße Rose von York" erlangte die königliche Wappenrose historische Berühmtheit während der Zeit der englischen Rosenkriege (s. S. 115).

Rosa × damascena

DAMASK ROSE

A TOUCH OF *ONE THOUSAND AND ONE NIGHTS*

Probably it's because of the intoxicatingly thick fragrance, like in a fairy tale from *One Thousand and One Nights*, or maybe because some varieties come into a second bloom in late summer. Either way, during the thirteenth century the Crusaders seem to have been so taken with *Rosa × damascena* that they brought it back to Europe. Particularly in Turkey, Iran, and Bulgaria, the damascene rose is still cultivated on a large scale for the production of rose oil.

DAMASZENER-ROSE

EIN HAUCH VON *TAUSENDUNDEINE NACHT*

Wahrscheinlich liegt es am betörend schweren Duft, der an ein Märchen denken lässt – vielleicht aber auch daran, dass manche Sorten im Spätsommer zur zweiten Blüte kommen: Auf jeden Fall schien es die *Rosa damascena* den Kreuzrittern so angetan zu haben, dass sie sie im 13. Jahrhundert nach Europa mitbrachten. Zur Rosenölgewinnung wird die Damaszener-Rose auch heute noch in großem Stil angebaut, vor allem in der Türkei, im Iran und in Bulgarien.

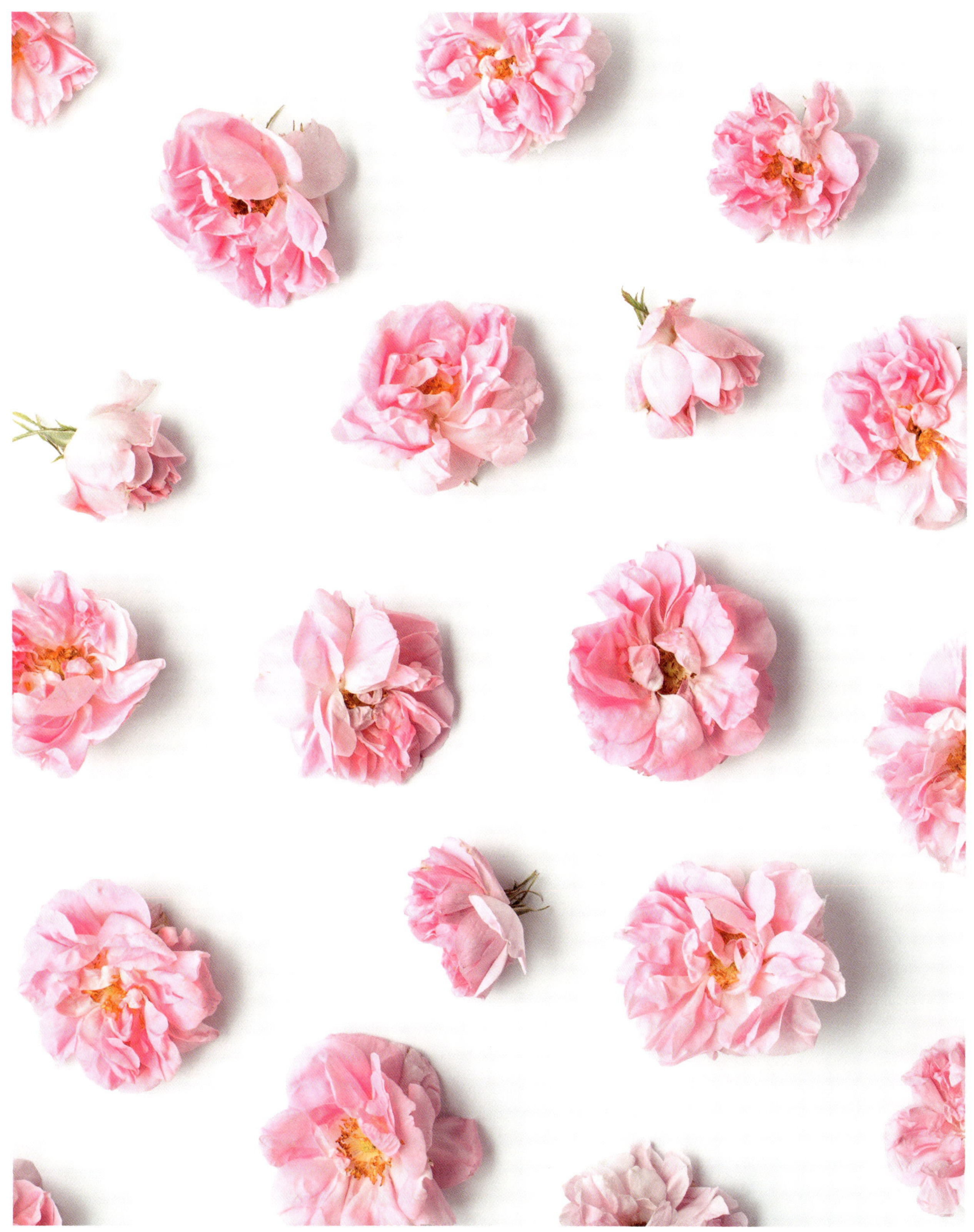

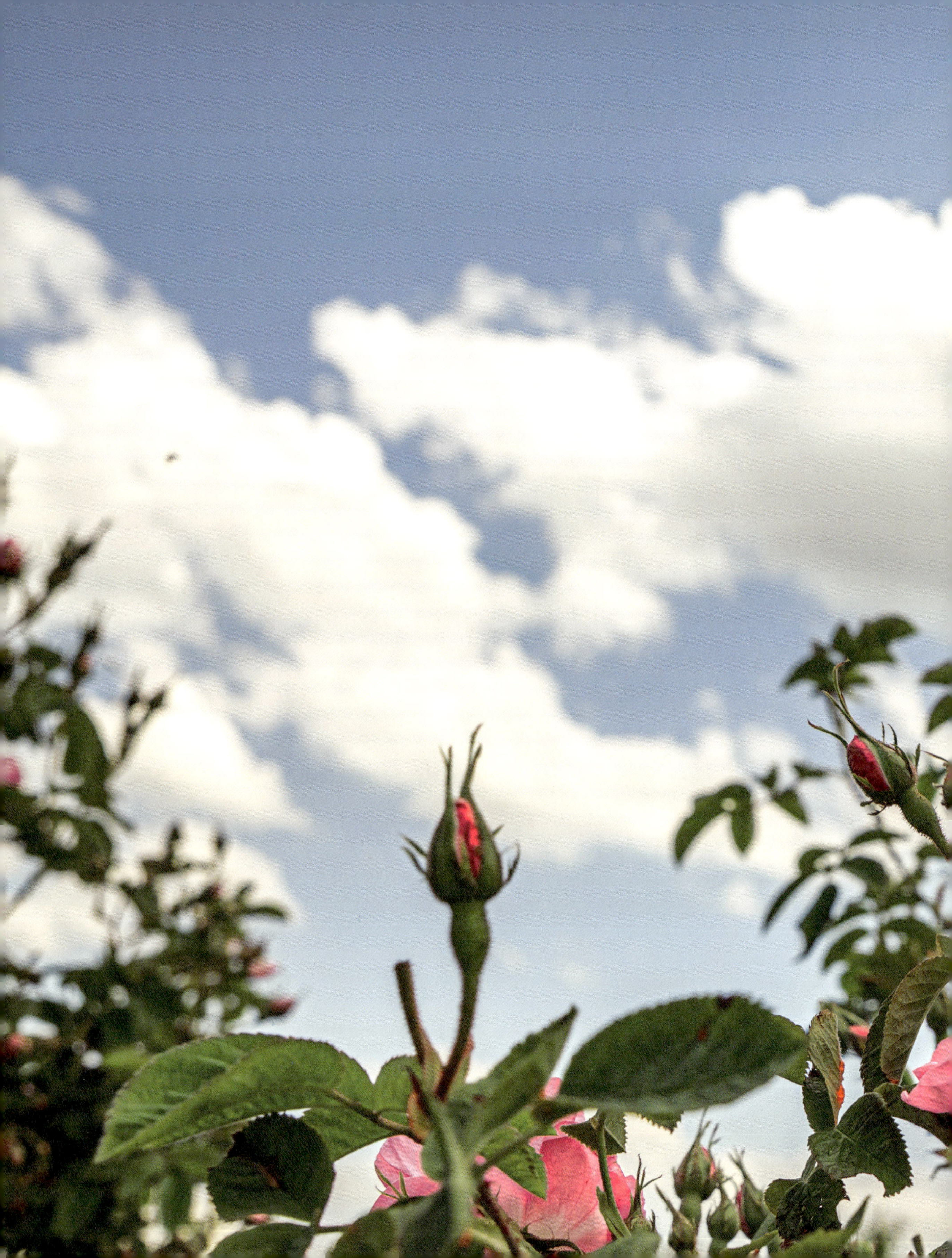

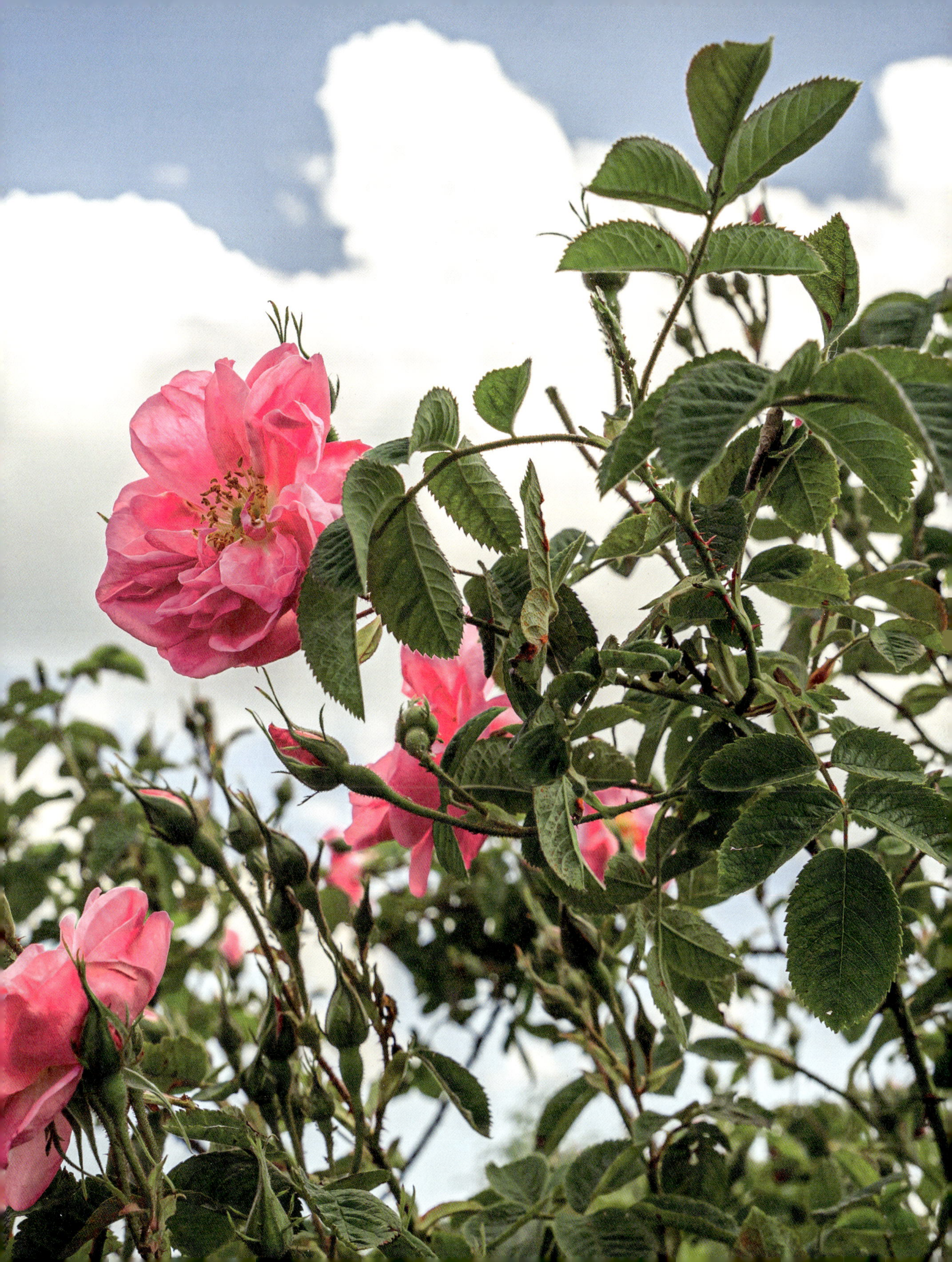

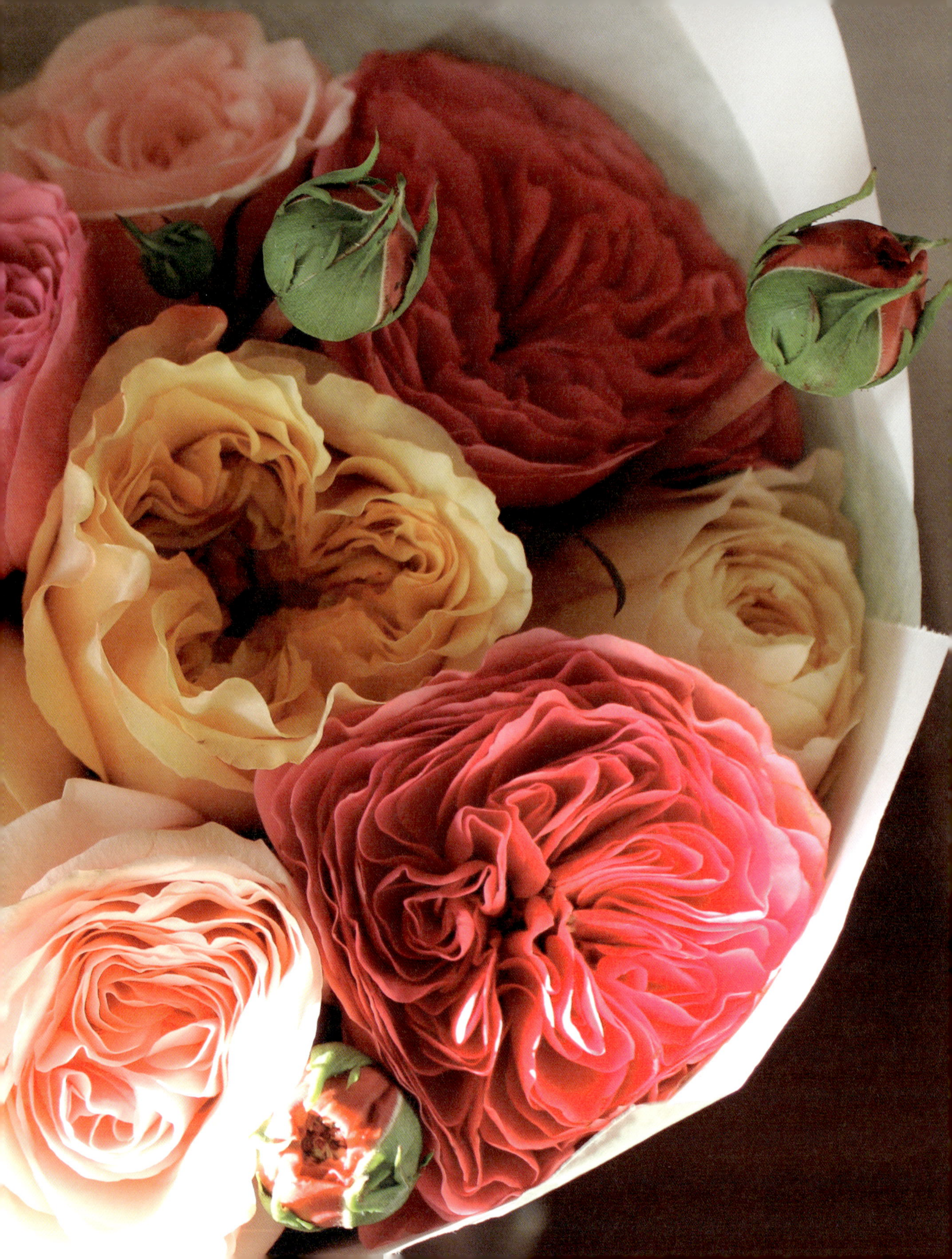

Rosa × centifolia

CENTIFOLIA

THE "HUNDRED LEAVED" PAINTER'S ROSE

Rosa centifolia is familiar from famous Flemish floral still lifes of the seventeenth century. Its lush, multilayered flowers made it extremely popular with the Dutch masters and is therefore also called the "painter's" rose. Since the spherical flower heads of many varieties of the hundred-petaled rose are even rain-resistant, they were once often found in cottage gardens. Probably thanks to its tightly nested petals, it also acquired the more prosaic nickname "cabbage rose".

ZENTIFOLIEN

DIE HUNDERTBLÄTTRIGEN MALERROSEN

Wir kennen die *Rosa centifolia* von den berühmten flämischen Blumenstillleben des 17. Jahrhunderts: Wegen ihrer üppigen, vielschichtigen Blüten war sie bei den niederländischen Meistern äußerst beliebt und wird deshalb auch Malerrose genannt. Da die kugeligen Blütenköpfe vieler Varietäten der Hundertblättrigen Rose sogar regenfest sind, fand man sie einst häufig in Bauerngärten. Vermutlich hat dank der dicht aneinandergeschmiegten Blütenblätter dort auch der prosaische Beiname Kohlrose seinen Ursprung.

Rosa × *centifolia ‚Muscosa‘*

MOSS-ROSE

A MAGICAL WHIM OF NATURE

The first "moss" roses appear near the end of the seventeenth century as the result of a spontaneous mutation in the lush centifolia. The "moss" that forms around the buds and sepals, which gives *Rosa centifolia muscosa* a nostalgic charm, are starkly aromatic scent glands. With the first unfilled flowers, targeted breeding could begin. One characteristic of the lovely moss rose—even its many tiny prickles look delicate—is that it buds abundantly, making it a popular rose for the garden.

RIGHT AND NEXT PAGE |
Functional mutation: The
"moss" on the buds consists
of scent glands.

RECHTE UND NÄCHSTE SEITE |
Mutation mit Funktion:
Das „Moos" auf den Knospen
besteht aus Duftdrüsen.

Rosa × centifolia ‚Muscosa‘

MOOSROSE

EINE ZAUBERHAFTE LAUNE DER NATUR

Durch spontane Mutation von üppig gefüllten Zentifolien entstanden gegen Ende des 17. Jahrhunderts die ersten Moosrosen: Das „Moos" der Knospen und Kelchblätter, das der *Rosa centifolia muscosa* einen nostalgischen Charme verleiht, sind Duftdrüsen mit intensivem Aroma. Mit den ersten ungefüllten Blüten konnte dann gezielt gezüchtet werden. Die liebliche Moosrose, bei der sogar die vielen winzigen Stacheln zart wirken, zeichnet sich durch eine reiche Knospenbildung aus, was sie zur beliebten Gartenrose macht.

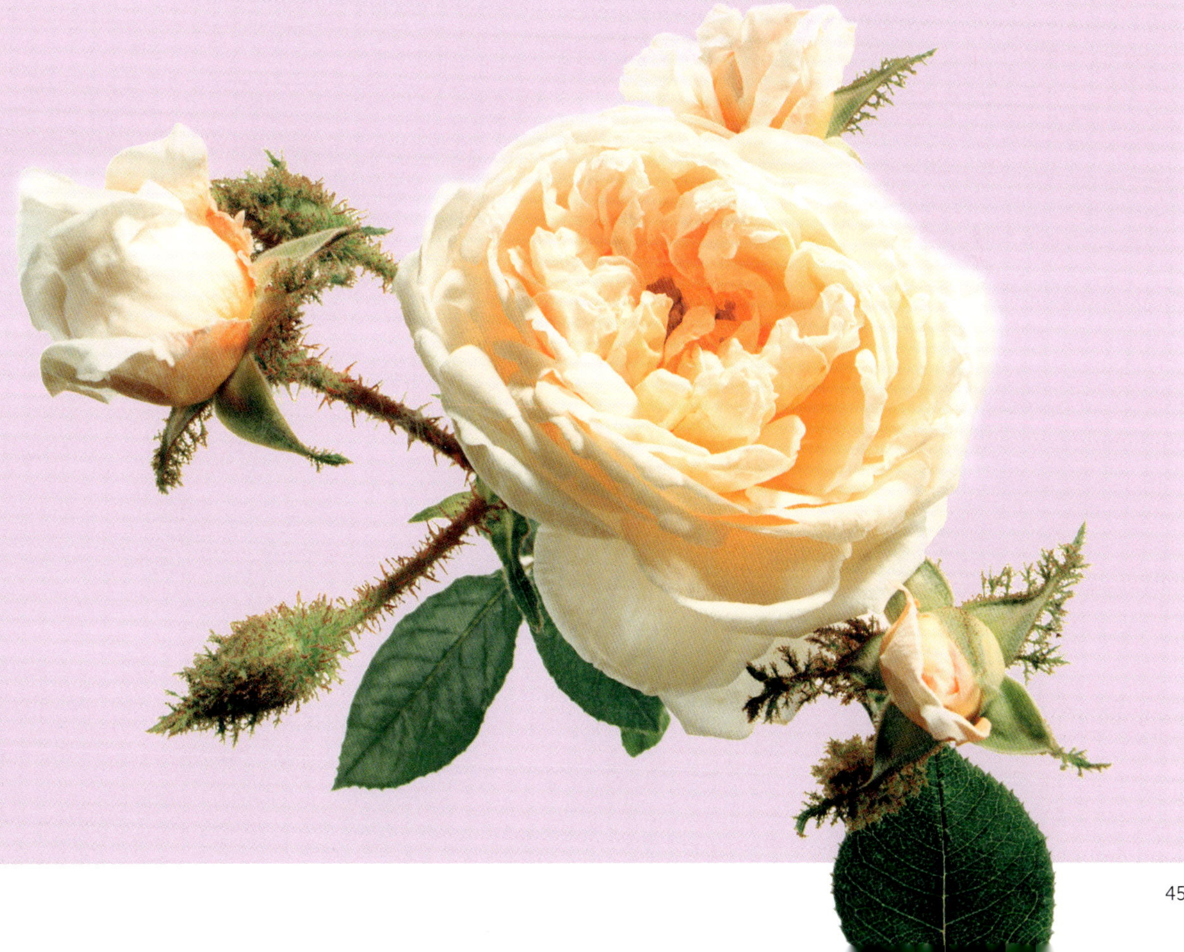

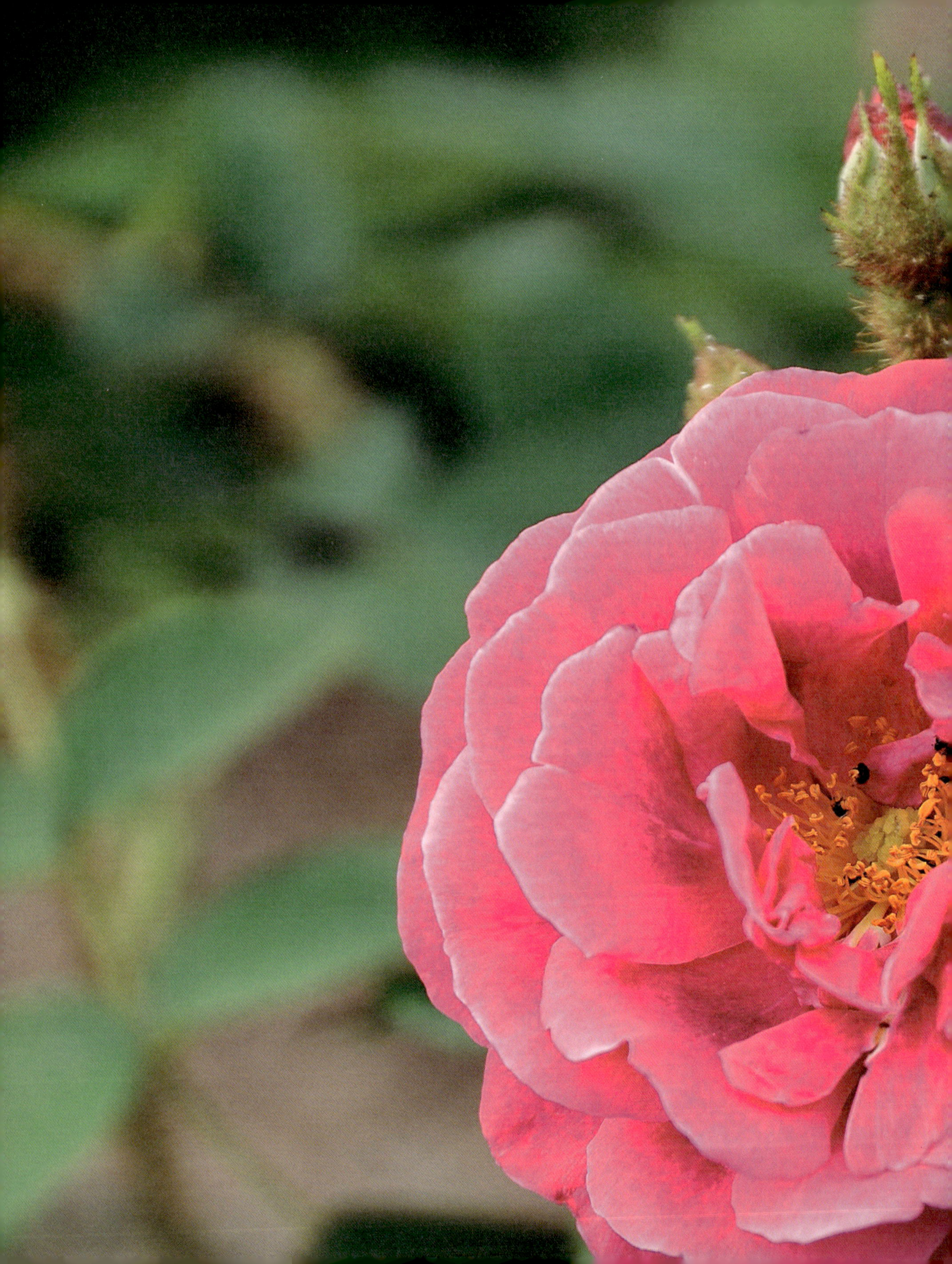

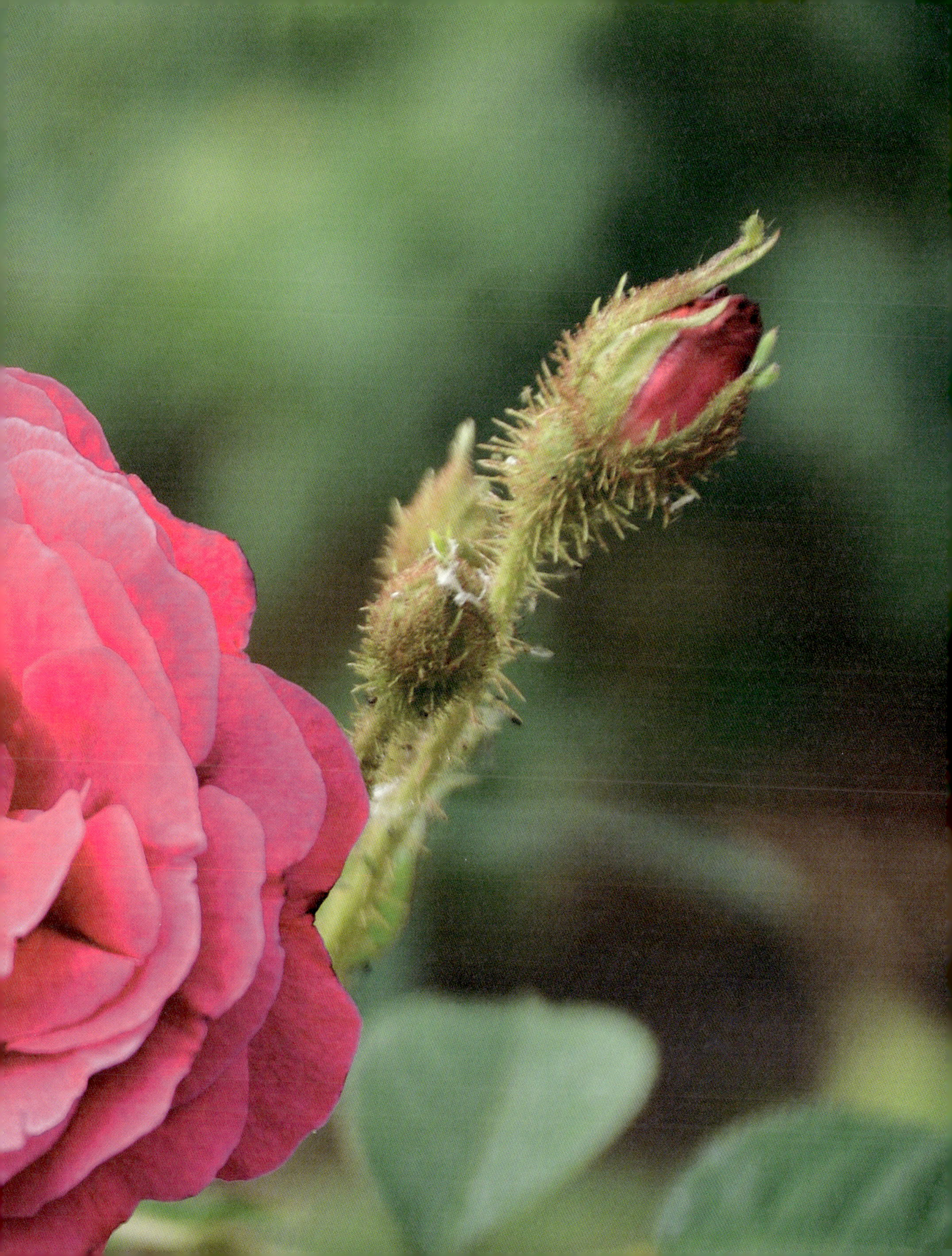

Rosa chinensis

CHINA ROSE

FAR-EASTERN BEAUTY BLOOMS AGAIN

The China rose is one of the oldest garden roses; its use as an ornamental was reported by Confucius already around 500 BCE. It found its way to Europe around 1800 via trade routes such as the Silk Road. It became popular in Europe for its ability to repeat-bloom and for its fine tea scent.

One particular freak of nature is the green-flowered *Rosa chinensis viridiflora*, a variety in which all parts of the flower have mutated into photosynthesizing green leaves (see picture on the right page). It is regarded as an unsightly curiosity.

CHINA-ROSE

WIEDERBLÜHENDE SCHÖNHEIT AUS FERNOST

Die China-Rose zählt zu den ältesten Gartenrosen – bereits um das Jahr 500 v. Chr. berichtete der chinesische Philosoph Konfuzius von ihrer Nutzung als Zierpflanze. Um das Jahr 1800 gelangte sie über Handelsrouten wie die Seidenstraße nach Europa, wo sie aufgrund ihrer Fähigkeit zur Nachblüte und ihres feinen Teedufts beliebt war.

Eine Laune der Natur ist die grünblütige *Rosa chinensis viridiflora*: Bei der gerne auch als „hässliche Kuriosität" bezeichneten Varietät sind alle Teile der Blüte zu grünen photosynthesefähigen Organen mutiert (im Bild rechts).

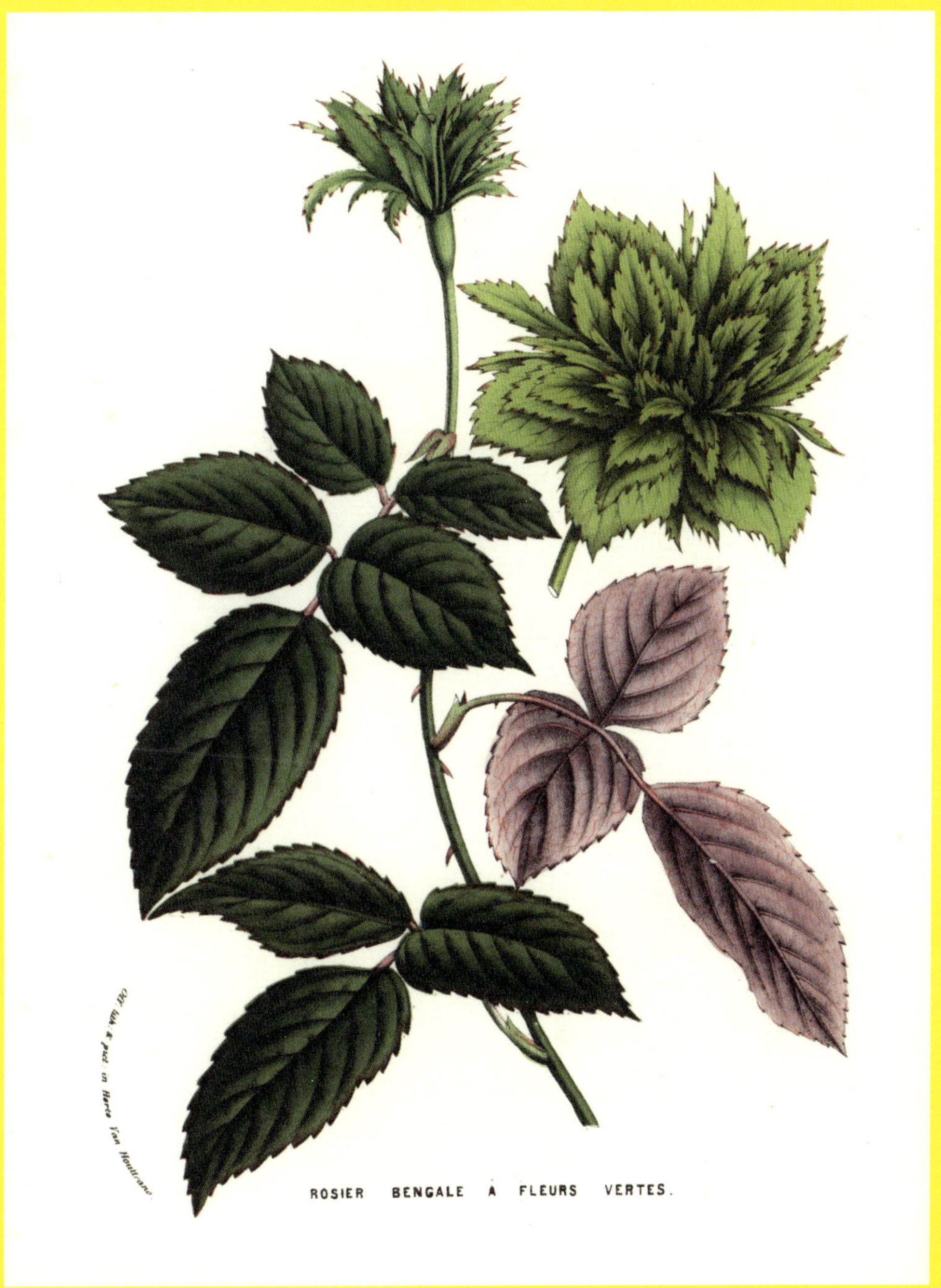
ROSIER BENGALE À FLEURS VERTES.

Modern roses

VIVE LA FRANCE!

THE MODERN ROSE-BREEDING ERA DAWNS WITH THE FIRST HYBRID TEA

The history of hybrid teas began by a sheer coincidence. The breeder Jean-Baptiste Guillot, in 1867, discovered a plant in a bed seeded with tea-scented roses. He continued to cultivate this plant, which is believed to have been a cross between the Chinese Tea rose and the remontant European bourbon rose. With the pink-flowering 'La France', the first hybrid tea was born. From then on, it would mark the divide between old roses and modern breeder roses.

The hybrid tea, with its tangled, slightly rolled petals, seems to have well suited contemporary tastes, which in the face of industrialization and the turmoil of war had acquired a taint of romantic longing. The English breeder Henry Bennett continued the success story of hybrid teas, employing new breeding principles and cultivating his plants in heated glass buildings.

Their erect habit, with straight stems and large flowers that maintain their beauty for a long time, sealed the triumph of these noble new roses. While growers concentrated on flower coloration and size, the enchanting fragrance of hybrid teas slowly dwindled. Fortunately though, recent years have seen a reversal in this trend. The diversity of color in today's varieties is huge, from white and baby pink, to hot pink, to many shades of red and sometimes even yellow roses.

Moderne Züchtungen

VIVE LA FRANCE!

DIE ENTSTEHUNG DER ERSTEN TEEHYBRIDE ERÖFFNET DAS KAPITEL DER MODERNEN ROSENZÜCHTUNG

Die Geschichte der Teehybriden begann durch einen schieren Zufall: 1867 entdeckte der Züchter Jean-Baptiste Guillot in einem Beet mit Teerosen-Saatgut eine Pflanze, die er weiter kultivierte. Man vermutet, dass sich in ihr chinesische Teerosen mit remontierenden europäischen Bourbonrosen vereinigt hatten. Mit der rosafarben blühenden *La France* war die erste Teehybride geboren, die fortan die Trennlinie zwischen alten Rosen und modernen Züchterrosen markieren sollte.

Offenbar traf die Teehybride mit ihren wirr angeordneten, leicht gerollten Blütenblättern den Geschmack der Zeit, in der angesichts von Industrialisierung und Kriegswirren romantisch angehauchte Sehnsüchte zutage traten. Der englische Züchter Henry Bennett schrieb mit neuen Zuchtprinzipien und dem Anbau in beheizten Glashäusern die Erfolgsgeschichte der Teehybriden fort.

Der aufrechte Wuchs an geraden Stängeln und große Blüten, die lange ihre Schönheit erhalten, sicherten den Siegeszug der neuen Edelrosen. Da die Züchter ihr Hauptaugenmerk auf Färbung und Größe der Blüten legten, verloren die Teehybriden langsam ihren zauberhaften Duft – ein Trend, der in den letzten Jahren zum Glück eine Wende erfuhr. Die Farbvielfalt der heutigen Sorten ist groß, von Weiß über Rosa, Pink und viele Rottöne bis hin zu gelbblütigen Rosen.

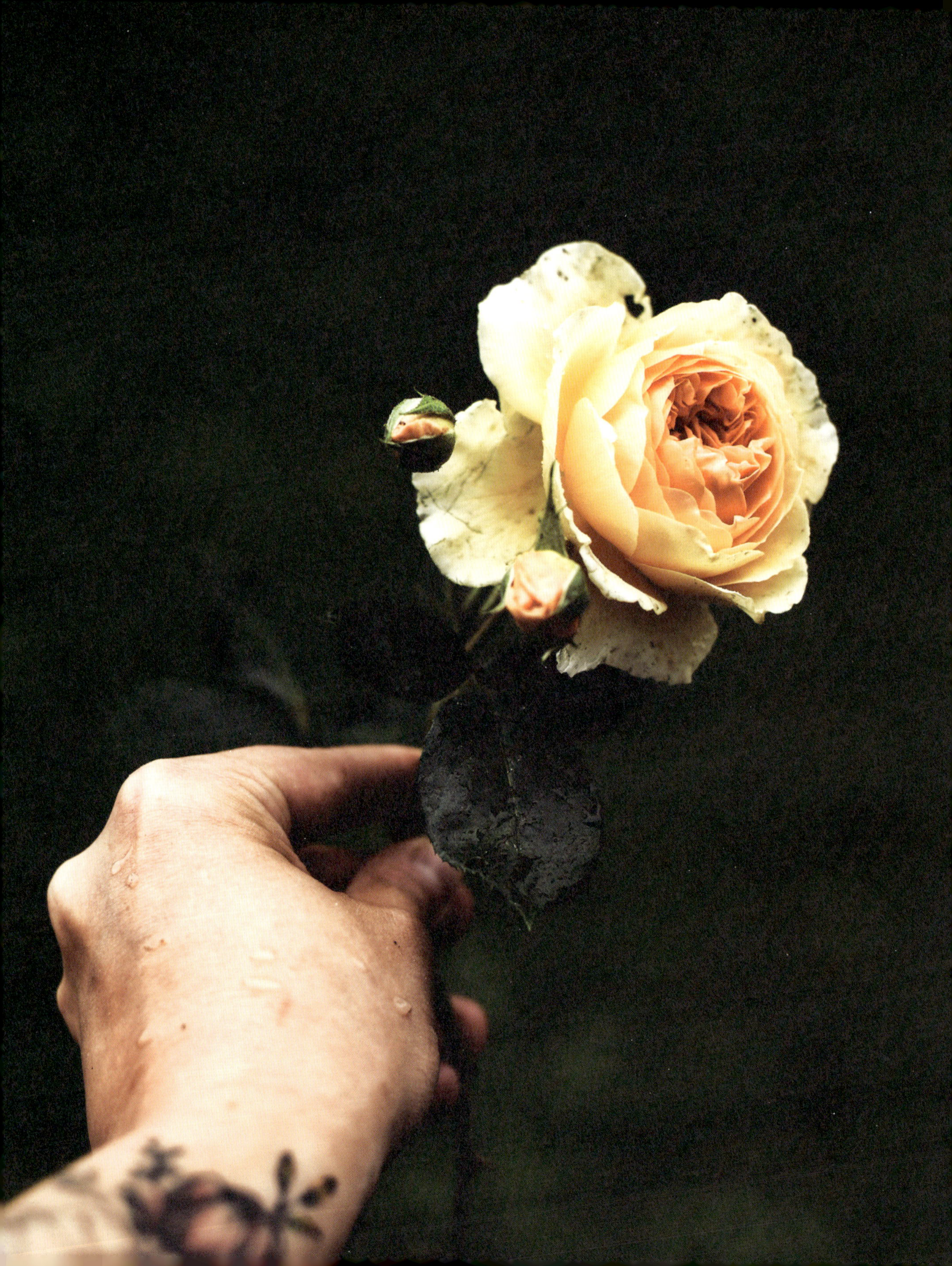

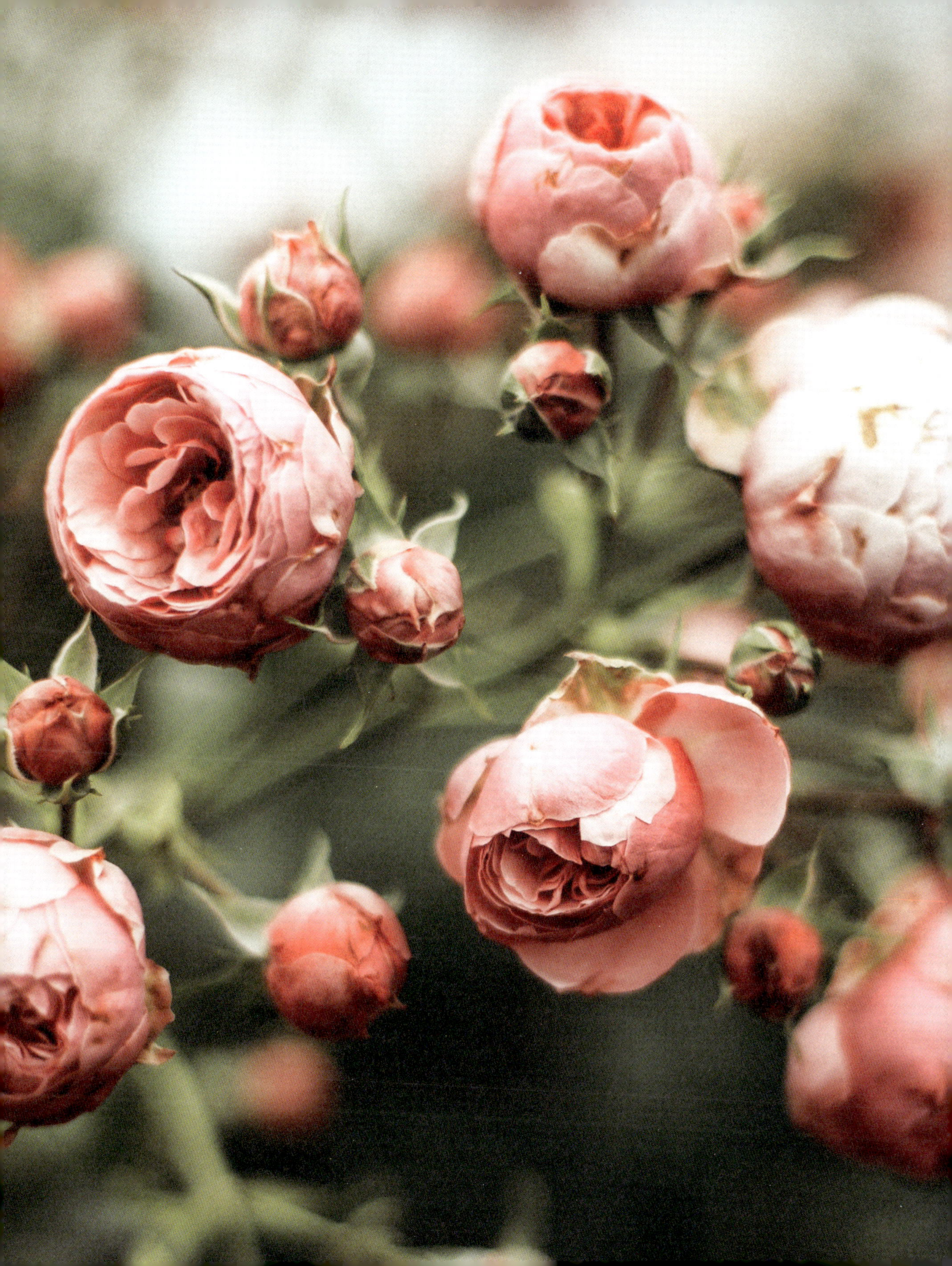

Modern roses

THE FATHER OF THE ENGLISH ROSES

DAVID AUSTIN'S SEARCH FOR THE IDEAL ROSE LED TO A RENAISSANCE OF FILL AND FRAGRANCE

Even as a teenager, David Austin, the son of an English farmer, was fascinated by roses. Inspired by his father's friend, a lupine grower, Austin tried his own hand at growing roses. In modern hybrid teas he valued only the color variety and their ability to repeat-blossom over the course of a single season. Instead, what he most cherished were old roses with their lush blossoms and beguiling fragrances. So, why not combine the best of both worlds? In 1961, Austin's first creation, the pink-flowered 'Constance Spry', was born.

Rose gardeners, however, entirely focused on hybrid teas, showed no interest in the new breed. This prompted Austin to take distribution into his own hands. Working out of his kitchen at first, he ultimately set up his own company, in 1969, and coined for his breeds the term "English rose". It has since become a household name among rose-lovers all over the world. Austin's breakthrough in the rose world came in 1983, when the fragrant, lush blooms of his yellow 'Graham Thomas' and pink 'Mary Rose' made a splash at the prestigious Chelsea Flower Show. In the years that followed, the wages this farmer's son from Shropshire reaped would be twenty-four Chelsea gold medals and a medal of the Order of the British Empire.

RIGHT PAGE | David Austin's creation 'Queen of Sweden'
NEXT PAGE | 'Sweet Juliet' and 'Lady Emma Hamilton'

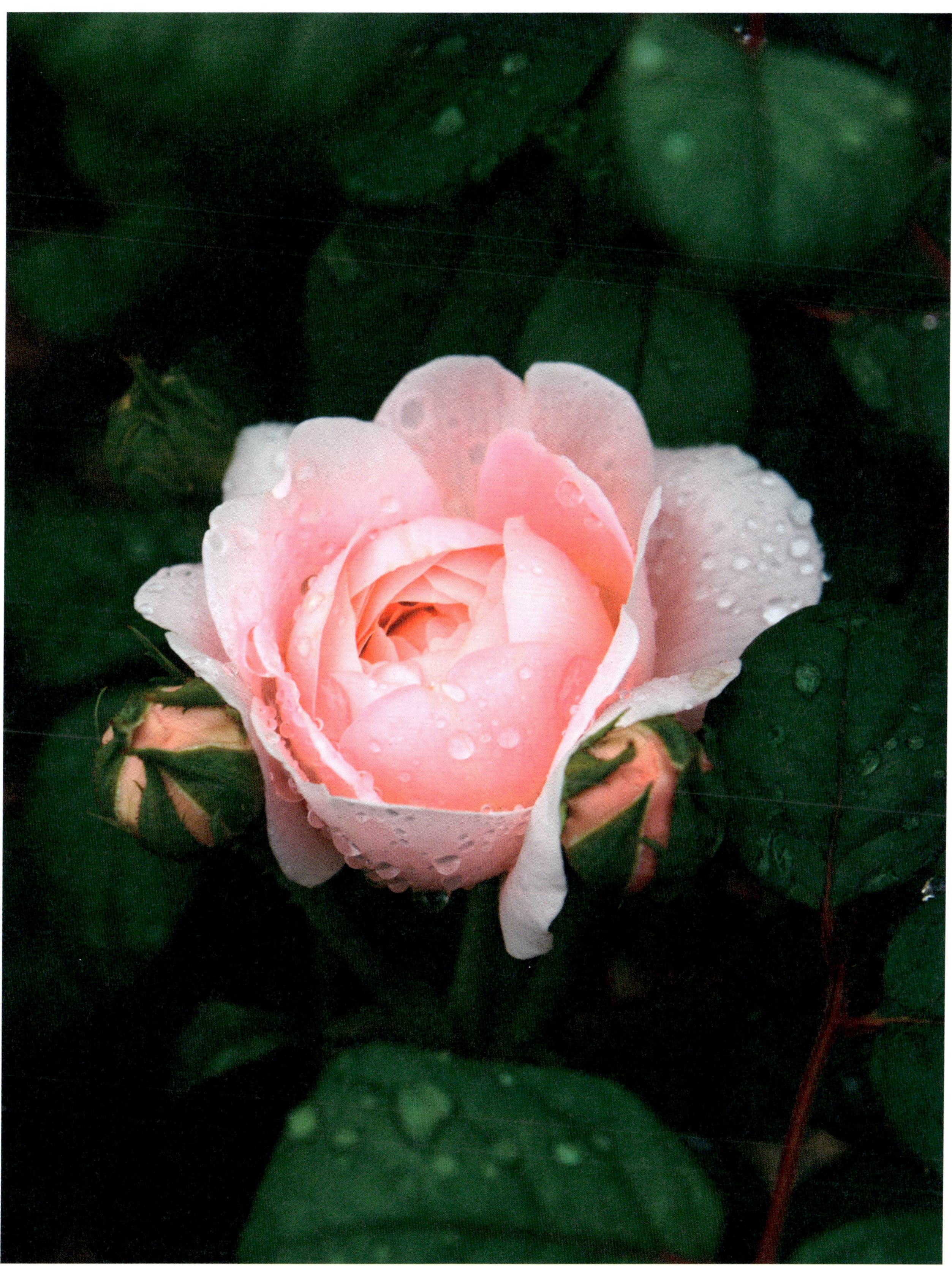

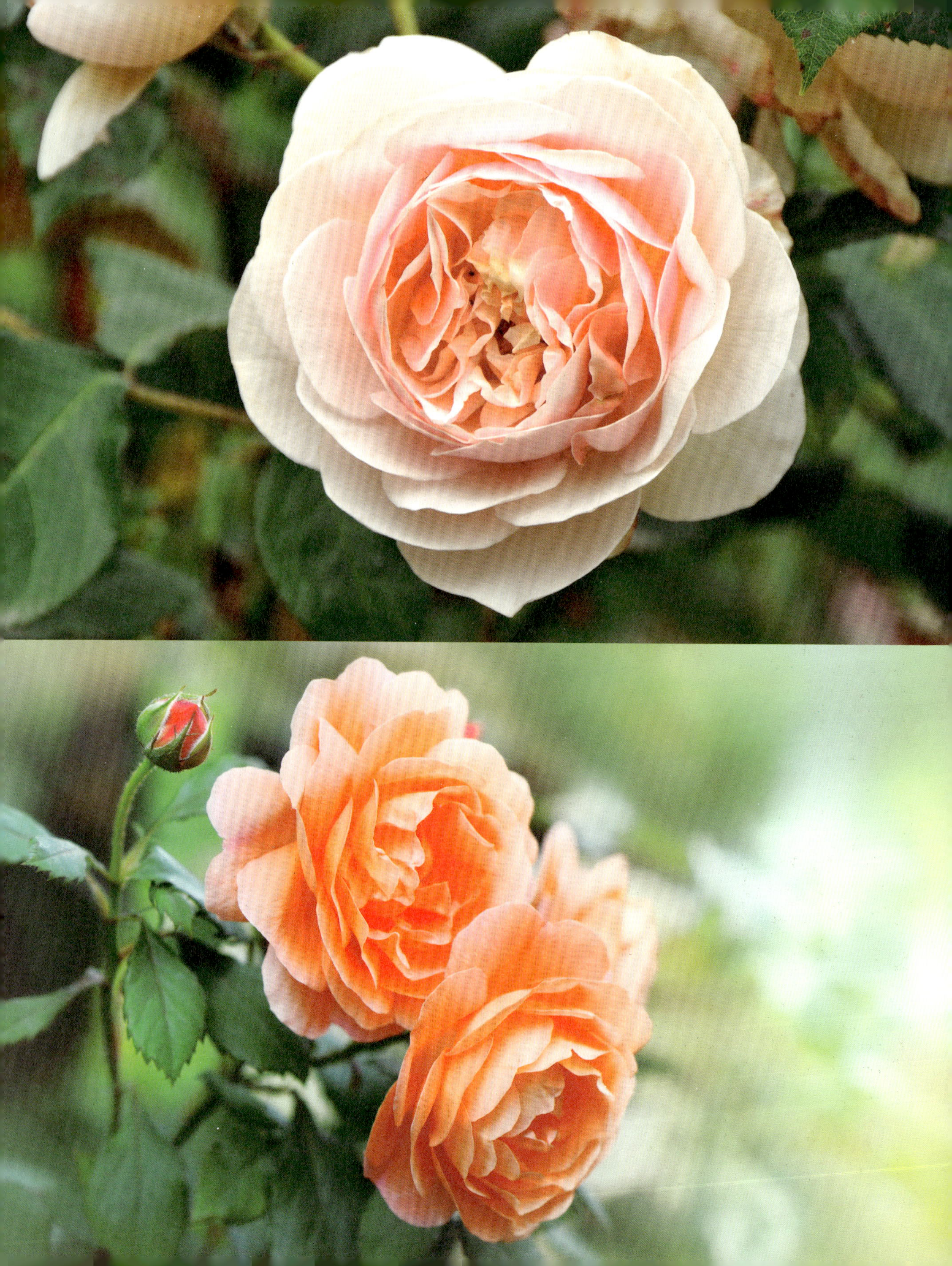

Moderne Züchtungen

DER VATER DER ENGLISCHEN ROSEN

DAVID AUSTINS SUCHE NACH DER IDEALEN ROSE FÜHRTE ZU EINER RENAISSANCE VON DUFT UND FÜLLE

Schon als Jugendlicher war der englische Bauernsohn David Austin von Rosen fasziniert. Inspiriert durch einen Freund seines Vaters, der Lupinen züchtete, versuchte sich Austin selbst als Rosenzüchter. An modernen Teehybriden schätzte er nur die Farbvielfalt und die Fähigkeit zu remontieren, also im Saisonverlauf wieder neu zur Blüte zu kommen. Seine große Liebe aber gehörte den alten Rosen mit ihren üppigen Blüten und dem betörenden Duft. Warum also nicht das Beste aus beiden Welten kombinieren? So entstand 1961 Austins erste Kreation, die rosafarben blühende *Constance Spry*.

Doch die Rosengärtner, ganz auf Teehybriden fokussiert, zeigten kein Interesse an der Neuzüchtung. So nahm Austin den Vertrieb selbst in die Hand – vom heimischen Küchentisch aus! 1969 gründete er schließlich sein eigenes Unternehmen und prägte für seine Züchtungen den Begriff „Englische Rose", der inzwischen bei Rosenliebhabern rund um den Globus in aller Munde ist. Austins Durchbruch in der Rosenwelt kam 1983, als die gelb blühende *Graham Thomas* und die rosafarbene *Mary Rose* mit ihren duftenden, üppigen Blüten bei der renommierten Chelsea Flower Show für Furore sorgten – in den folgenden Jahren sollten 24 Chelsea-Goldmedaillen und der Orden des Britischen Empire der Lohn für den Bauernsohn aus Shropshire sein.

VORHERIGE SEITE | David Austins Kreation *Queen of Sweden*

LINKE SEITE | *Sweet Juliet* und *Lady Emma Hamilton*

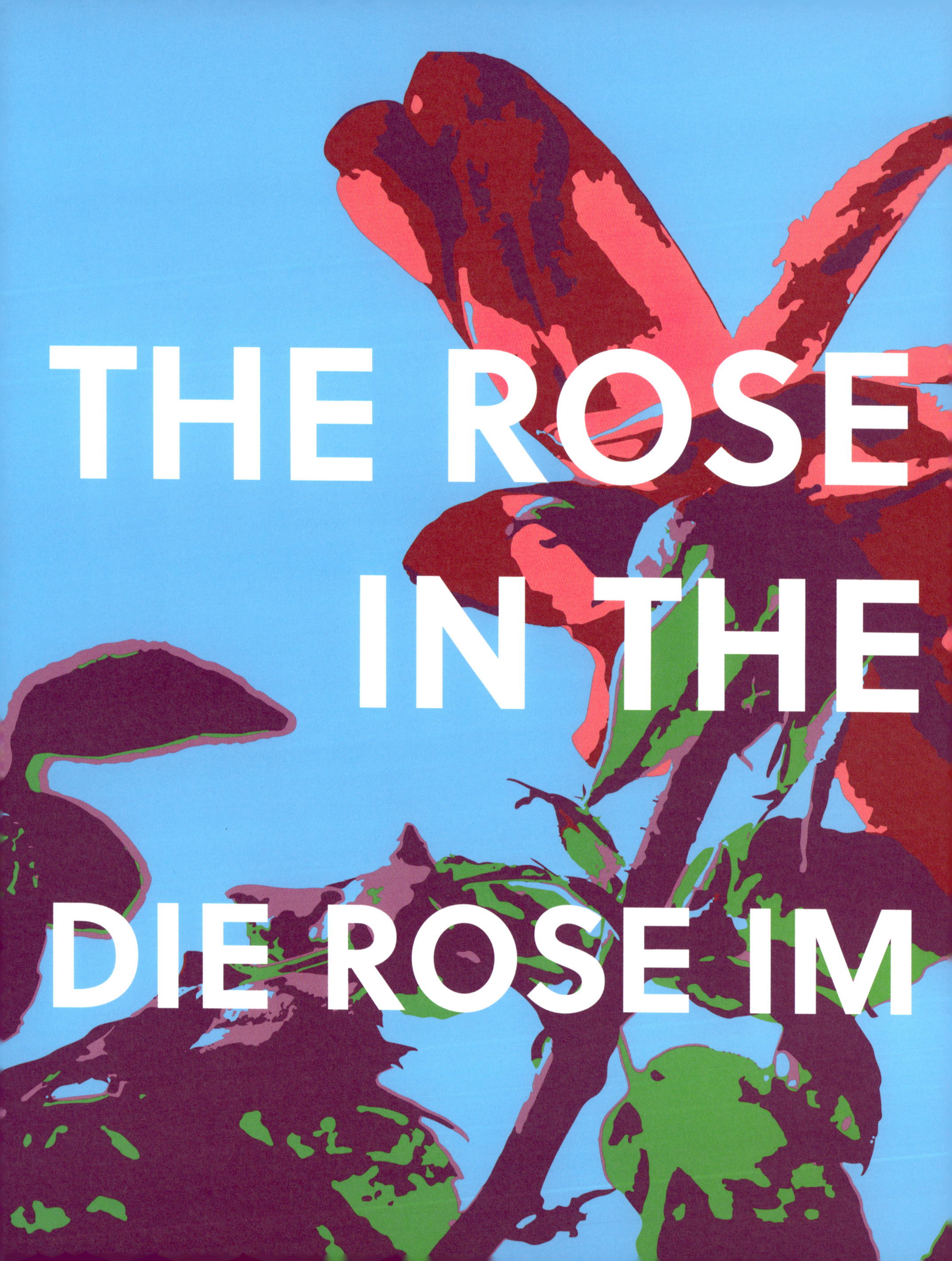

THE ROSE
IN THE
DIE ROSE IM

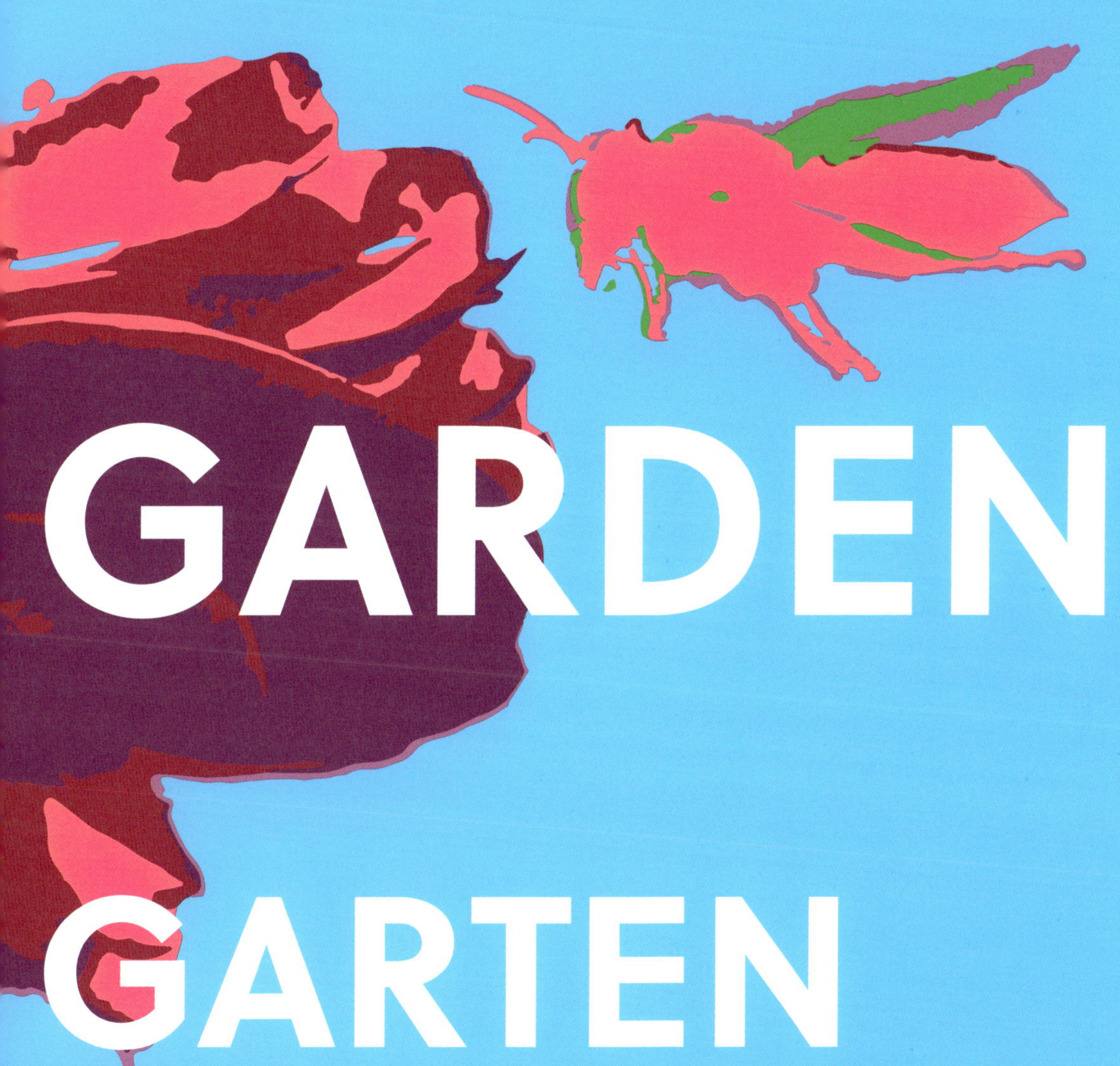

GARDEN

GARTEN

The rose conservatory at Sangerhausen

SHOW OF SUPERLATIVES

WITH 80,000 ROSES, EUROPE'S LARGEST ROSE COLLECTION BLOOMS IN THE SOUTHERN HARZ

The rose capital of Europe is Sangerhausen, in the German state of Saxony-Anhalt. Thanks to the Verein Deutscher Rosenfreunde (Association of German rosists), this old mining town has become Europe's rosy center. As early as 1898, dedicated rose-lovers and breeders had entertained the notion of having their own rosarium, a place to collect old, endangered rose species and get a handle on a great many new varieties. The breeders probably also wanted to counteract the dubious practice of reintroducing already-existing rose varieties under new names.

Europe's largest rose collection opened its doors in 1903. It had taken five years to plant 2,800 bushes (comprising 1,100 rose varieties) in and around Sangerhausen's old town park. Many had been donated by rose collector Albert Hoffmann, who had called upon rose lovers all over the world to send rare varieties; and so, what today forms the core of the Deutsche Genbank Rose (German genetic repository for roses) was created in no small part thanks to him.

Today more than 80,000 rose bushes comprising 8,700 rose varieties and species bloom in Sangerhausen. And it is not only during the primary blossom from May that rose lovers stroll, enraptured, through the varied parks of Europe's largest collection.

NEXT PAGE | Everything is where it belongs:
The rosarium in its day began tracking a bewildering
array of new cultivars, real or pretended.

Europa-Rosarium Sangerhausen

ROSENSCHAU DER SUPERLATIVE

MIT 80 000 ROSEN BLÜHT EUROPAS GRÖSSTE ROSENSAMMLUNG IM SÜDHARZ AUF

Europas Herz der Rosen liegt in Sachsen-Anhalt, genauer gesagt in Sangerhausen. Dass die alte Bergbaustadt zum europäischen Zentrum der Rosen wurde, ist dem Verein Deutscher Rosenfreunde zu verdanken. Bereits 1898 trugen sich engagierte Rosenliebhaber und -züchter mit der Idee, ein eigenes Rosarium zu gründen, um alte, vom Aussterben bedrohte Rosenarten zu sammeln und einen Überblick über die Vielfalt an Neuheiten zu erhalten. Die Züchter wollten damit wohl auch der Unsitte entgegenwirken, dass bereits existierende Rosensorten einfach wieder unter neuem Namen präsentiert wurden.

Im Jahr 1903 öffnete Europas größte Rosensammlung ihre Tore: Im und rund um den alten Sangerhauser Stadtpark waren binnen fünf Jahren 2800 Rosenstöcke von 1100 Rosensorten gepflanzt worden, vornehmlich gestiftet vom Rosensammler Albert Hoffmann. Auch dank dessen Aufruf an Rosenfreunde in aller Welt, Raritäten nach Sangerhausen zu schicken, entstand so das Herzstück der heutigen „Deutschen Genbank Rose".

Mittlerweile blühen in Sangerhausen über 80 000 Rosenstöcke aus 8700 Rosensorten und -arten, und nicht nur zur Hauptblütezeit ab Mai schlendern Rosenfreunde verzückt durch die abwechslungsreichen Parkanlagen. Auch geheiratet wird hier gern.

NÄCHSTE SEITE | Ordnungssinn: Das Rosarium wurde u. a. gegründet, um die damals zahlreichen Neuzüchtungen im Auge zu behalten – und diejenigen, die nur als solche deklariert wurden.

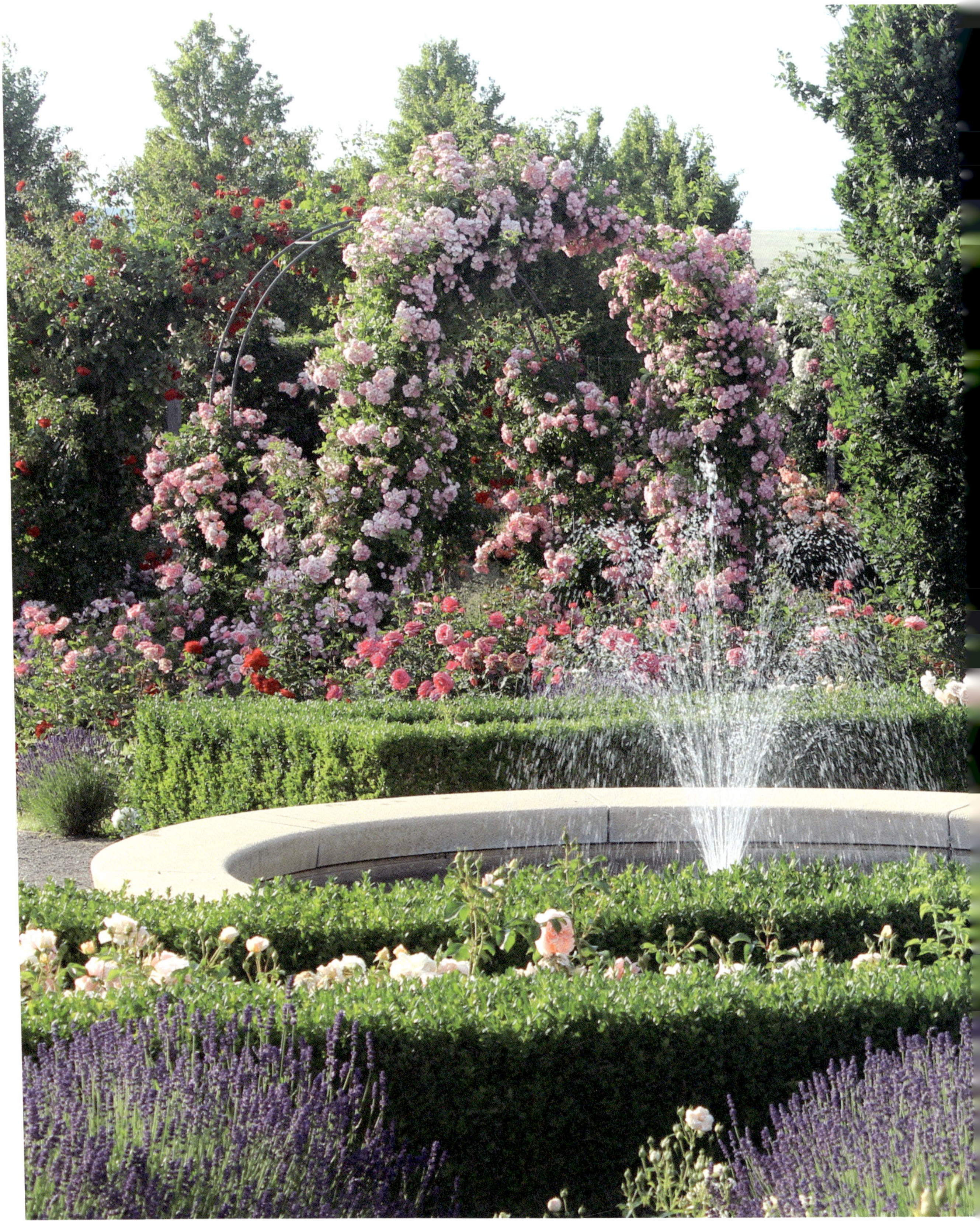

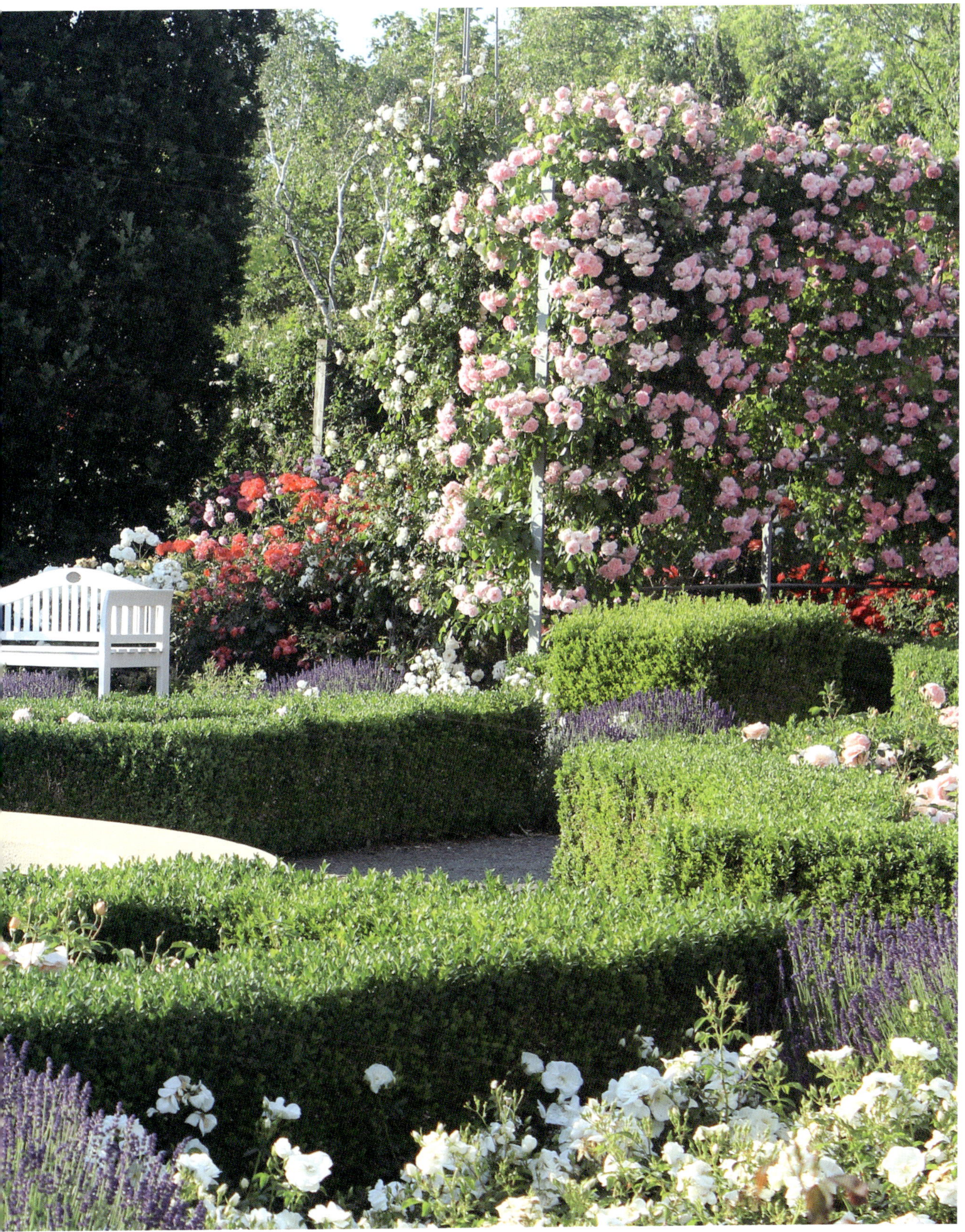

For your own garden

A FEAST FOR BEES

INSECT ARMAGEDDON PROMPTS A RE-EVALUATION AMONG ROSE-LOVERS

Visually, lush, full rose varieties are true gems of the art of breeding, and there is no question that they often smell wonderful. But today it is more important than ever that gardens and parks contain flowers that not only the eyes but also the bees and other pollinators can feast on. Roses with the popular lush, full blossoms are in many cases problematic, because the more petals develop within the flower, the less room there is for pollen-bearing stamens. So guess again if you thought they provided nourishment for bees and insects.

Fortunately, unfilled wild roses are now making a comeback and can contribute to counteracting the die-off of insects. Several breeders have seen the writing on the wall and are increasingly refocusing on unfilled or semi-filled rose varieties that offer insects open access to their stamens. Not only bees benefit from this trend, but also nature lovers, enthusiastic about unfilled and historic wild roses, appreciate them because of their magnificent rose hips, which set wonderful accents in the garden even in fall and winter.

NEXT PAGE | Bees and bumblebees alike gather nectar from flowering plants.

Für den eigenen Garten

EIN FEST FÜR BIENEN

IMMER MEHR INSEKTEN STERBEN AUS. AUCH ROSENFREUNDE BRINGT DAS ZUM UMDENKEN

Üppig gefüllte Rosensorten sind optisch wahre Prachtstücke der Züchterkunst und duften oft wunderbar, keine Frage. Doch heute ist es wichtiger denn je, dass Blumen in Gärten und Parks nicht nur eine Augenweide sind, sondern auch als „Insektenweide" für Bienen & Co und somit als Nahrungsquelle zur Verfügung stehen. Das Problem mit den beliebten üppig gefüllten Blüten: Bei vielen dieser Rosensorten wurden auch im Inneren der Blüten immer mehr Blütenblätter herangezüchtet, auf Kosten der pollentragenden Staubgefäße. Nahrung für Bienen und andere Insekten? Fehlanzeige!

Doch zum Glück erleben die ungefüllten Wildrosen inzwischen ein Comeback und können so ihren Beitrag leisten, um dem Insektensterben entgegenzuwirken. Etliche Züchter haben die Zeichen der Zeit erkannt und konzentrieren sich vermehrt wieder auf ungefüllte oder halb gefüllte Rosensorten, die Insekten freien Zugang zu ihren Staubgefäßen bieten. Und nicht nur die Bienen profitieren von diesem Trend: Naturliebhaber, die sich für ungefüllte Rosen und historische Wildrosen begeistern, schätzen diese nicht zuletzt wegen der prachtvollen Hagebutten, die im Garten auch im Herbst und Winter wunderbare Akzente setzen.

NÄCHSTE SEITE | Nicht nur Honigbienen, sondern auch Wildbienen wie Hummeln sammeln den Nektar blühender Pflanzen ein.

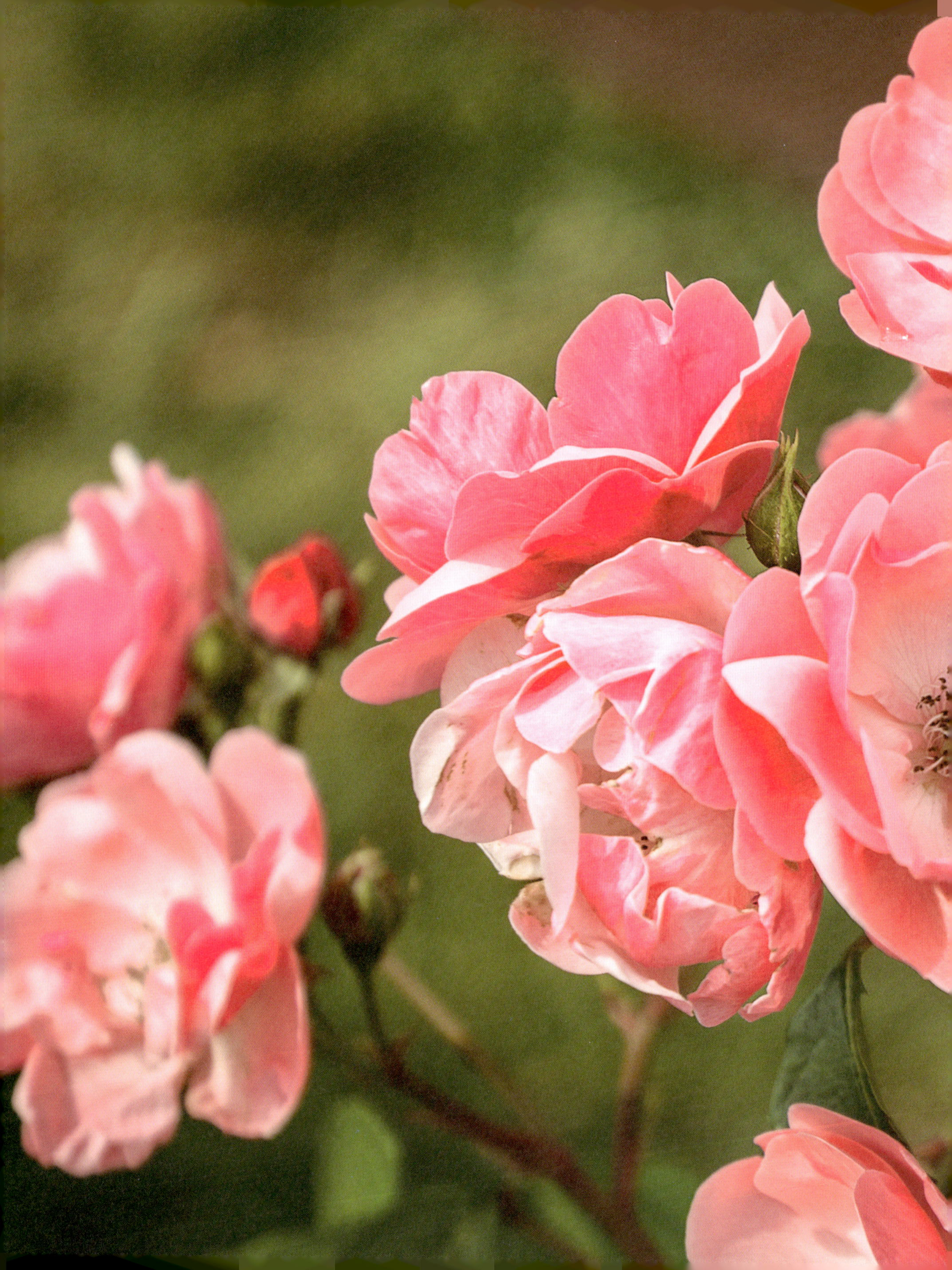

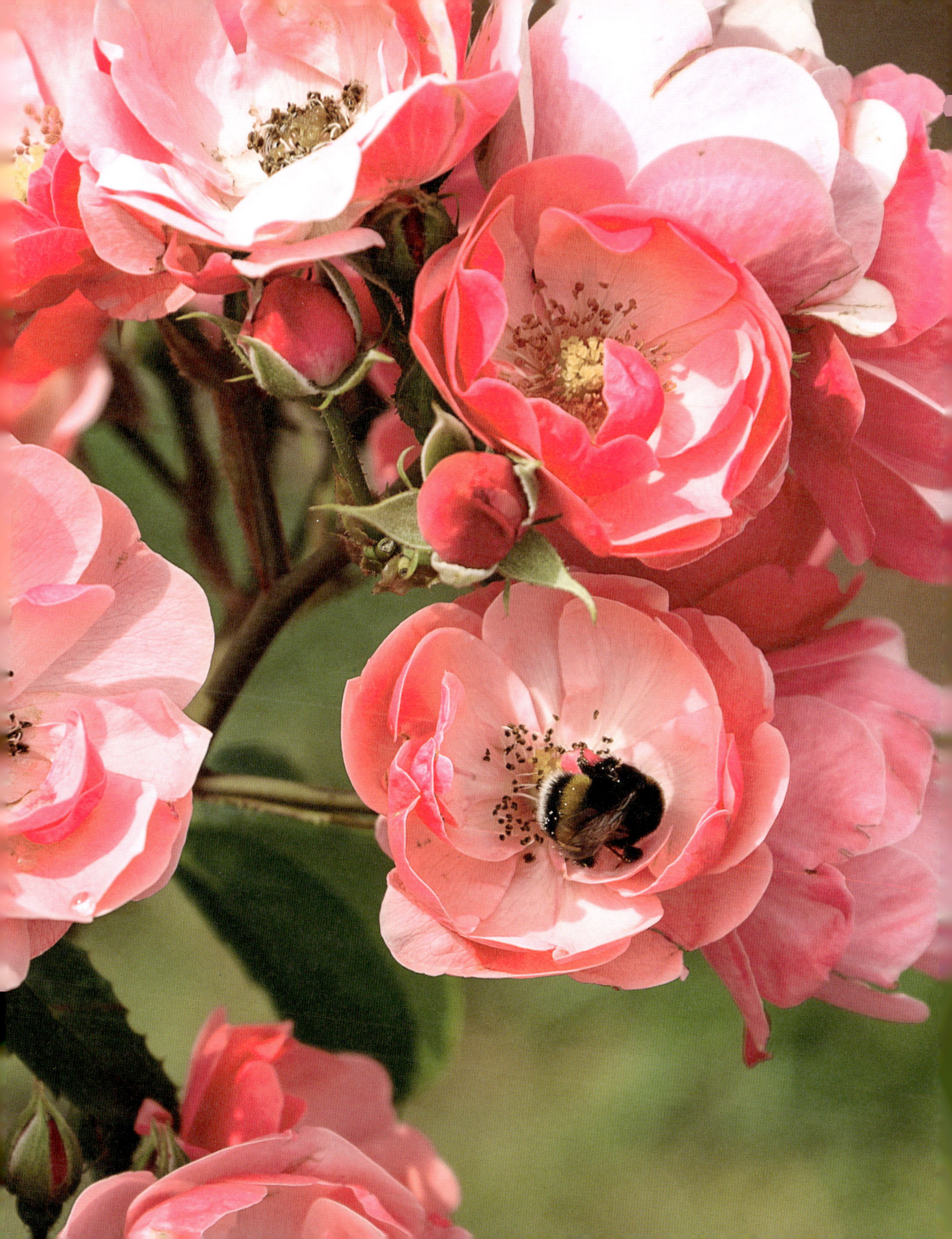

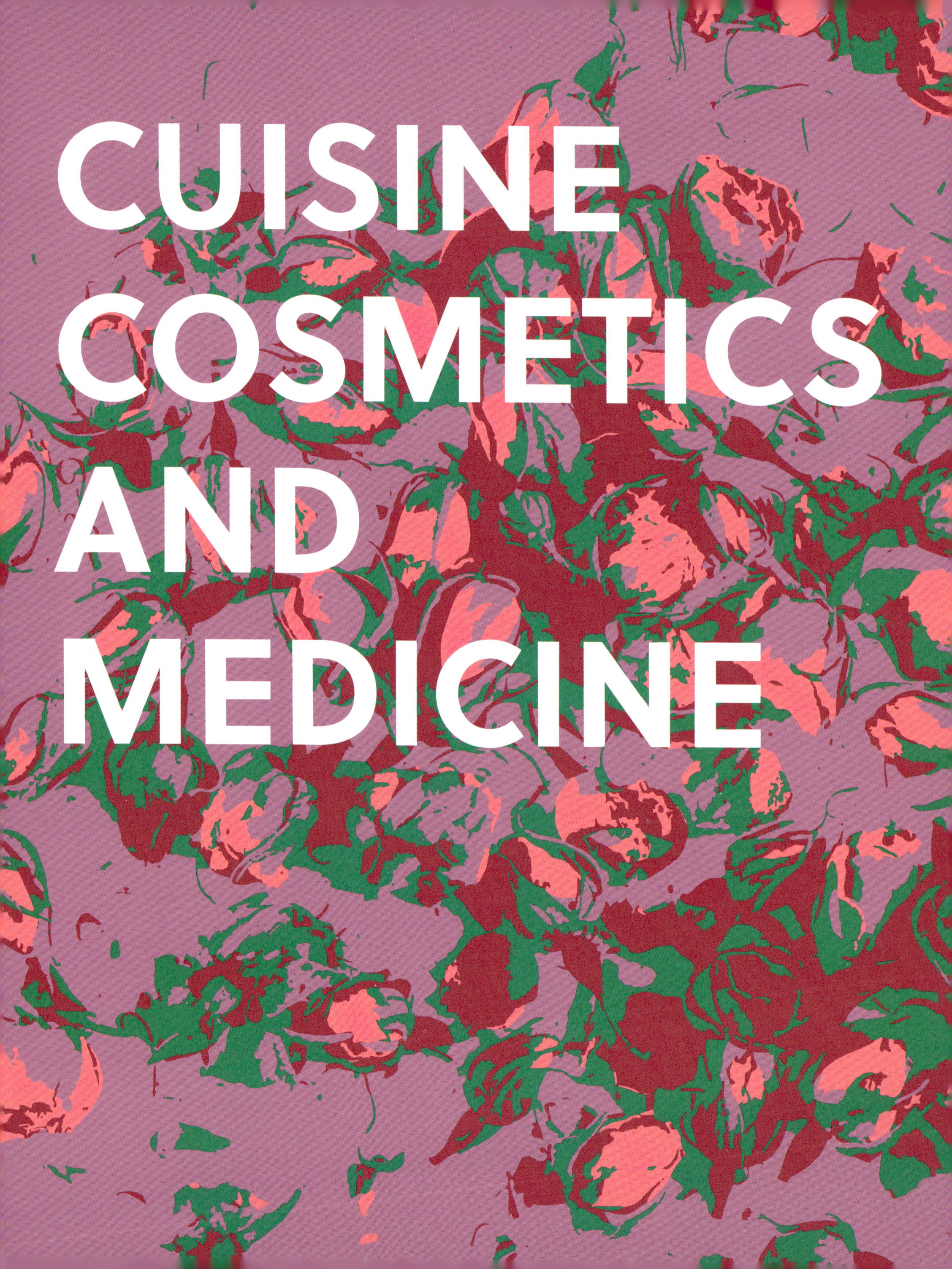
CUISINE
COSMETICS
AND
MEDICINE

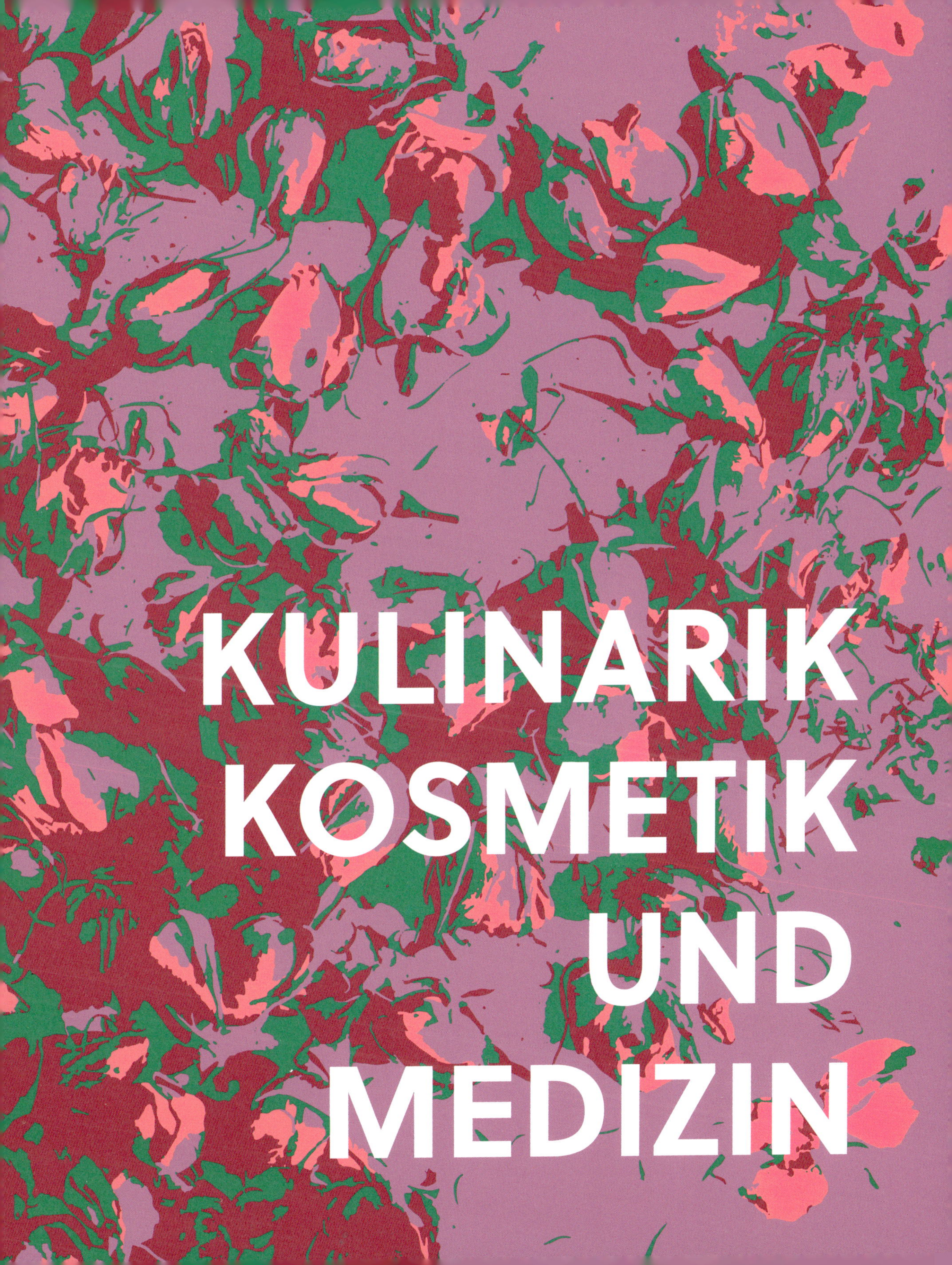
KULINARIK
KOSMETIK
UND
MEDIZIN

Rose petals, rose water etc.

EXQUISITE FLAVOR

ITS SEDUCTIVE AROMA MAKES THE ROSE
A SOUGHT-AFTER CULINARY INGREDIENT

Rose blossoms have a long history as a Lucullan extravagance. Ancient Romans in high society would indulge in the luxury of aromatizing wine with rose petals.

But many recipes for roses originated in herbal medicine, and in bygone times they would have been taken not so much with pleasure but rather for their medicinal properties. A rose tea, for example, is said to alleviate cramps and soothe pain. Roses today are often used to flavor black and green teas. The fruit of the rose, too, has found its way into teapots throughout history: vitamin-rich rose hips have always been regarded as a refreshing boost to the immune system.

Coming down from the seventeenth century is the *Kräuter-Buch* (Book of herbs), an herbal by Iacobus T. Tabernaemontanus, with recipes for rose vinegar and honey and the like, including one for a thirst-quenching, refreshing rose julep. Sugar or honey is an excellent means of preserving the aroma of the rose, and so it is no surprise that sweet rose recipes are popular among the epicurean-minded: rose blossom jelly at breakfast; a refreshing shot of rose syrup in prosecco; rose flowers in a fine liqueur; a magical-seeming decoration on a cake. Its preferred form in Middle Eastern cuisines is rose water, derived through the distillation of rose oil—just think about some oh-so-seductive confectioneries, like rose lokum or rose water baklava.

NEXT PAGE | Rose jelly (left) is made from (among other things) Damask roses or centifolias and is often flavored with a touch of lavender. Both as an opulent decoration or with flavors such as rose water: Roses do double duty in festive confectionery (right).

Rosenblüten, Rosenwasser & Co

EXQUISITER GESCHMACK

IHR VERFÜHRERISCHES AROMA MACHT DIE ROSE ZUR BEGEHRTEN ZUTAT IN DER KÜCHE

Rosenblüten als lukullische Extravaganz haben eine lange Geschichte: Schon im antiken Rom gönnte man sich in der feinen Gesellschaft den Luxus, Wein mit Rosenblättern zu aromatisieren.

Doch viele Rezepte mit Rosen entstammen der Kräuterheilkunde und waren einst mehr Medizin als Genussmittel: So soll Rosentee krampflösend und schmerzlindernd wirken. Heute werden mit Rosenblüten gerne Schwarz- und Grüntees aromatisiert. Auch die Frucht der Rose findet sich seit jeher in der Teekanne wieder – die vitaminreichen Hagebutten galten schon immer als abwehrstärkend und erfrischend.

Aus dem 17. Jahrhundert ist uns das *Kräuter-Buch* von Jacobus T. Tabernaemontanus mit Rezepten von Rosenessig über Rosenhonig bis hin zu Rosen-Julep als erfrischendem Durstlöscher überliefert. Da sich mit Zucker oder Honig das Aroma der Rose wunderbar konservieren lässt, verwundert es nicht, dass süße Rosenrezepte auch heute bei Genießern beliebt sind: Rosenblütengelee zum Frühstück, ein erfrischender Schuss Rosensirup im Prosecco sowie Rosenblüten in feinem Likör oder als zauberhafter Tortenschmuck. In der orientalischen Küche findet bevorzugt das bei der Destillation von Rosenöl gewonnene Rosenwasser Verwendung – man denke nur an so verführerische Süßigkeiten wie Rosen-Lokum oder Baklava mit Rosenwasser.

NÄCHSTE SEITE | Rosengelee (links) wird unter anderem aus Damaszener-Rosen oder Zentifolien gekocht und gern mit einem Hauch Lavendel verfeinert. Auf festlichen Torten (rechts) spielen Rosen gleich eine Doppelrolle – als opulente Dekoration und als Geschmacksträger, etwa in Form von Rosenwasser.

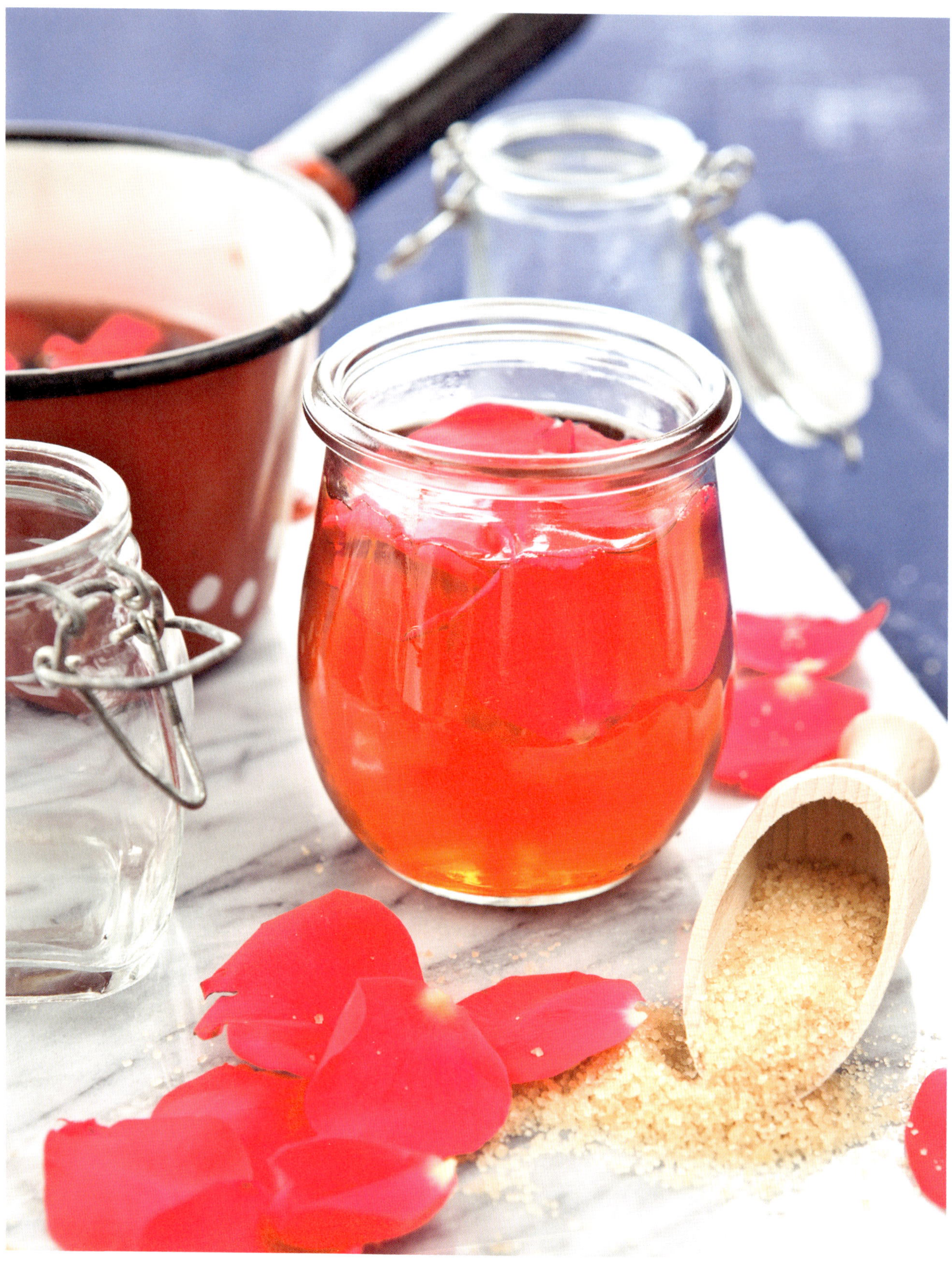

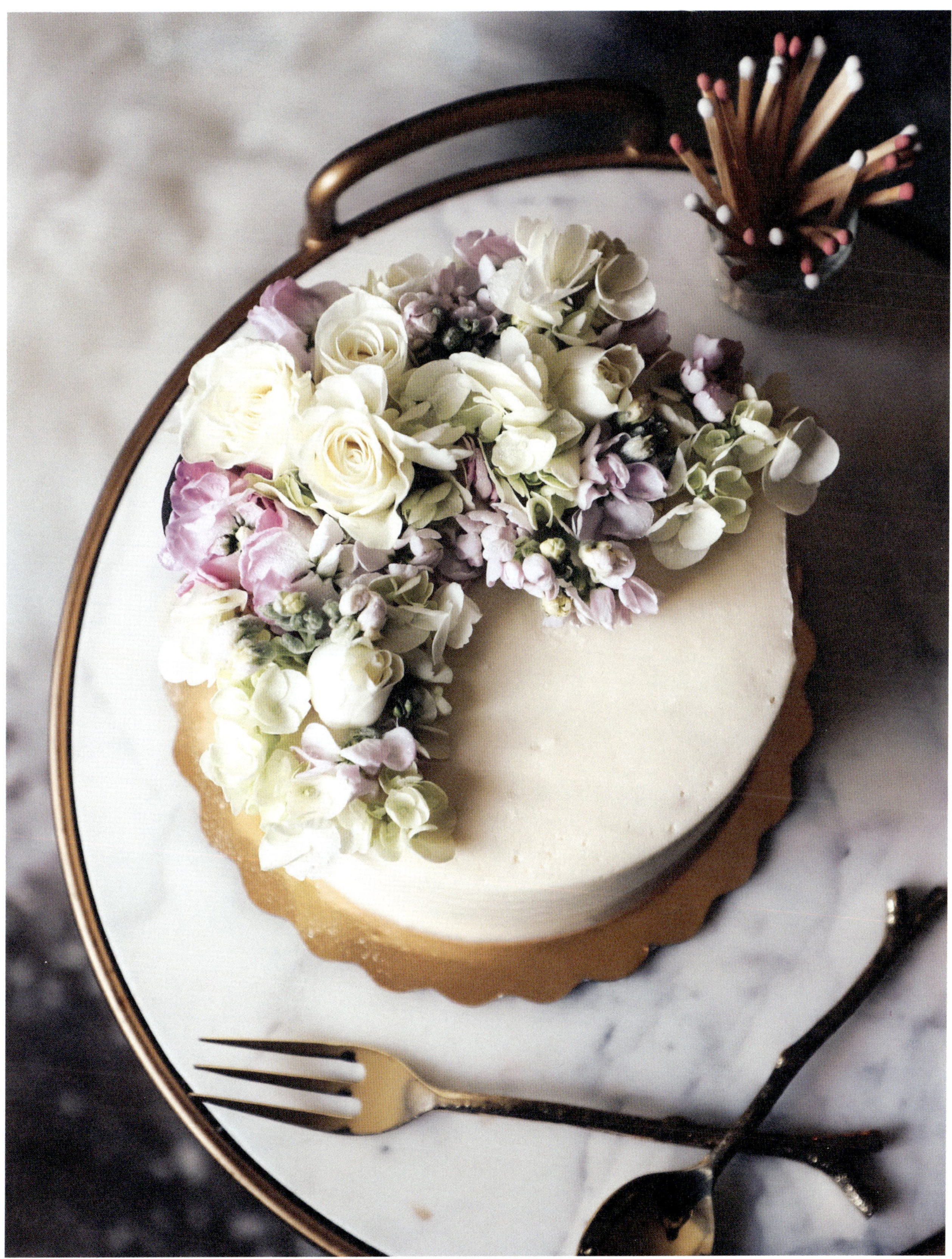

Perfumes & cosmetics

ENCHANTING AROMA

CRAFTING EXQUISITE PERFUMES FROM THE SEDUCTIVE AROMA OF ROSE BLOSSOMS IS AN ARTFORM

The damascene rose, *Rosa damascena* (p. 38), and the Provence or cabbage rose, *Rosa centifolia* (p. 42), are among the most intensely aromatic. Steam distillation is employed to extract rose oil from their flower petals in a process that consumes a huge quantity of roses, between three and five metric tons to produce a liter of the aromatic essential oil that is the essence of many high-end perfumes. Here, think classic fragrances like Chanel No 5. Good thing one of the by-products of this distillation is rose water, which is used not only in skin care but also as a culinary ingredient.

And by the way, the extraction of rose absolutes is even more cost-intensive: young rose buds are steeped in a solvent, usually one that contains alcohol, which is then separated out again by cooling, filtration, and steaming. This process produces a pasty *concrète* (concrete) that, mixed once more with alcohol, becomes *absolue*. Because the process uses less heat, *absolue* is supposedly closer in fragrance to the original rose.

Rose oil contains around four hundred different aromatic compounds, but what results in the fascinating scent of the rose is the balance: a complete natural compositional harmony that lacks nothing and which perfumers have always strived to achieve.

NEXT PAGE | Undiminished popularity: Rose notes in perfumes and cosmetics

Parfüm & Kosmetik

BETÖRENDER DUFT

DIE KUNST, AUS DEM VERFÜHRERISCHEN AROMA VON ROSENBLÜTEN EXQUISITE PARFÜMS ZU KREIEREN

Zu den am intensivsten duftenden Rosen zählen die Damaszener-Rose (s. S. 38) und die Zentifolien-Rose (s. S. 42). Aus deren Blütenblättern wird mithilfe von Wasserdampfdestillation Rosenöl gewonnen, und zwar auf ziemlich aufwendige Art: Zwischen 3000 und 5000 Kilogramm Rosenblüten sind nötig, um einen Liter des duftenden ätherischen Öls herzustellen, das die Essenz vieler hochwertiger Parfüms ist – man denke hier an den Duftklassiker Chanel No 5. Gut, dass als Nebenprodukt der Destillation auch Rosenwasser entsteht, das beispielsweise in der Hautpflege, aber auch in der Küche Verwendung findet.

Noch kostenintensiver ist übrigens die Gewinnung von Rose Absolue: Hier werden junge Rosenknospen in ein meist alkoholisches Lösungsmittel eingelegt, das anschließend durch Kühlung, Filtration und Verdampfung aufgelöst wird. Daraus entsteht das pastöse Concrète, das, wiederum mit Alkohol versetzt, zum Absolue wird. Da hier weniger Hitze aufgewendet wird, soll das Absolue noch näher am Duft der ursprünglichen Blüte sein.

Im Rosenöl sind rund 400 verschiedene Aromakomponenten vorhanden – und erst aus der Balance dieser Elemente ergibt sich der betörende Duft der Rose: eine vollkommene Kompositionsharmonie der Natur, die Parfümeure seit jeher anstreben.

NÄCHSTE SEITE | Bis heute ein Hit:
Rosennoten in Parfüms und Kosmetika

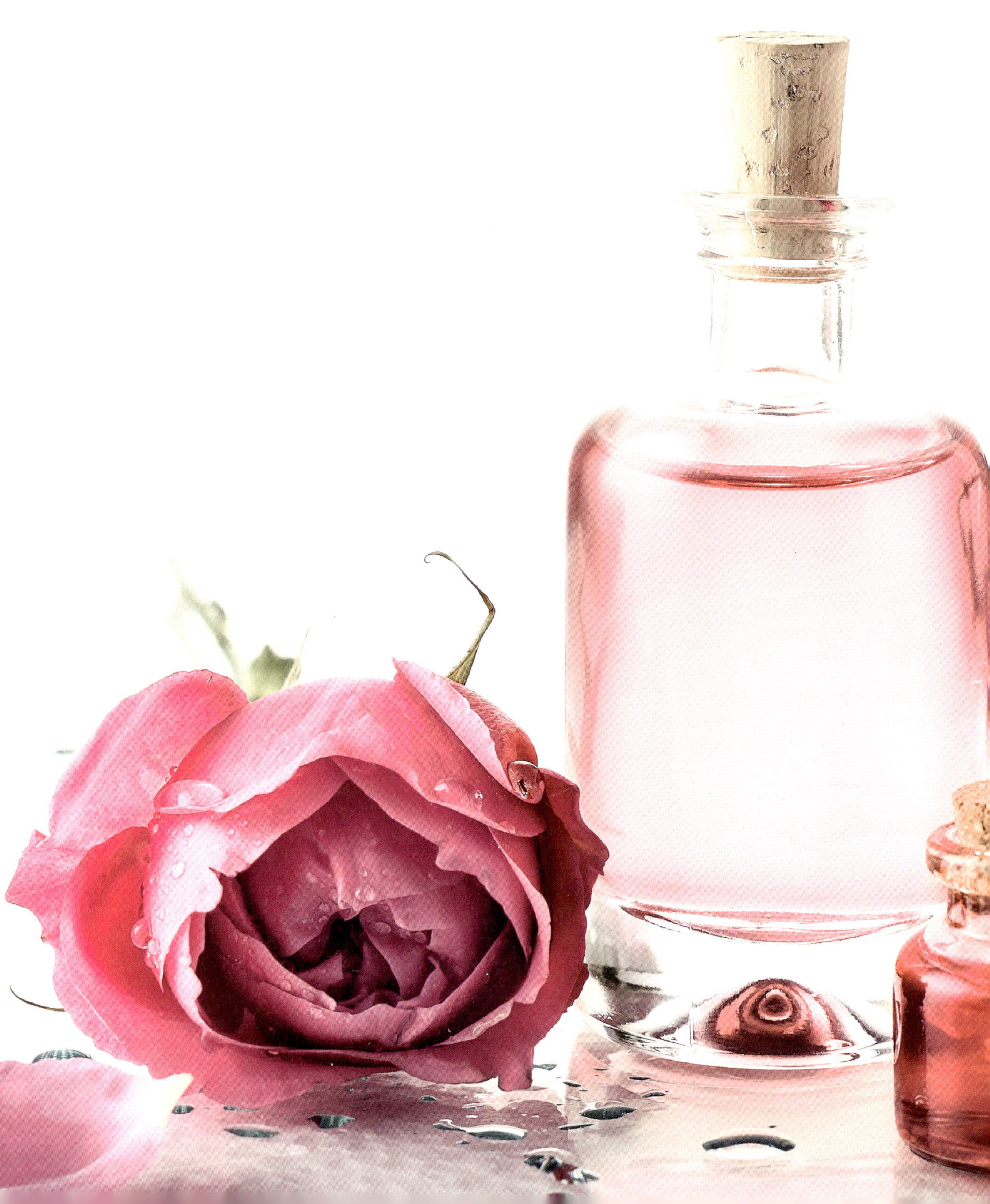

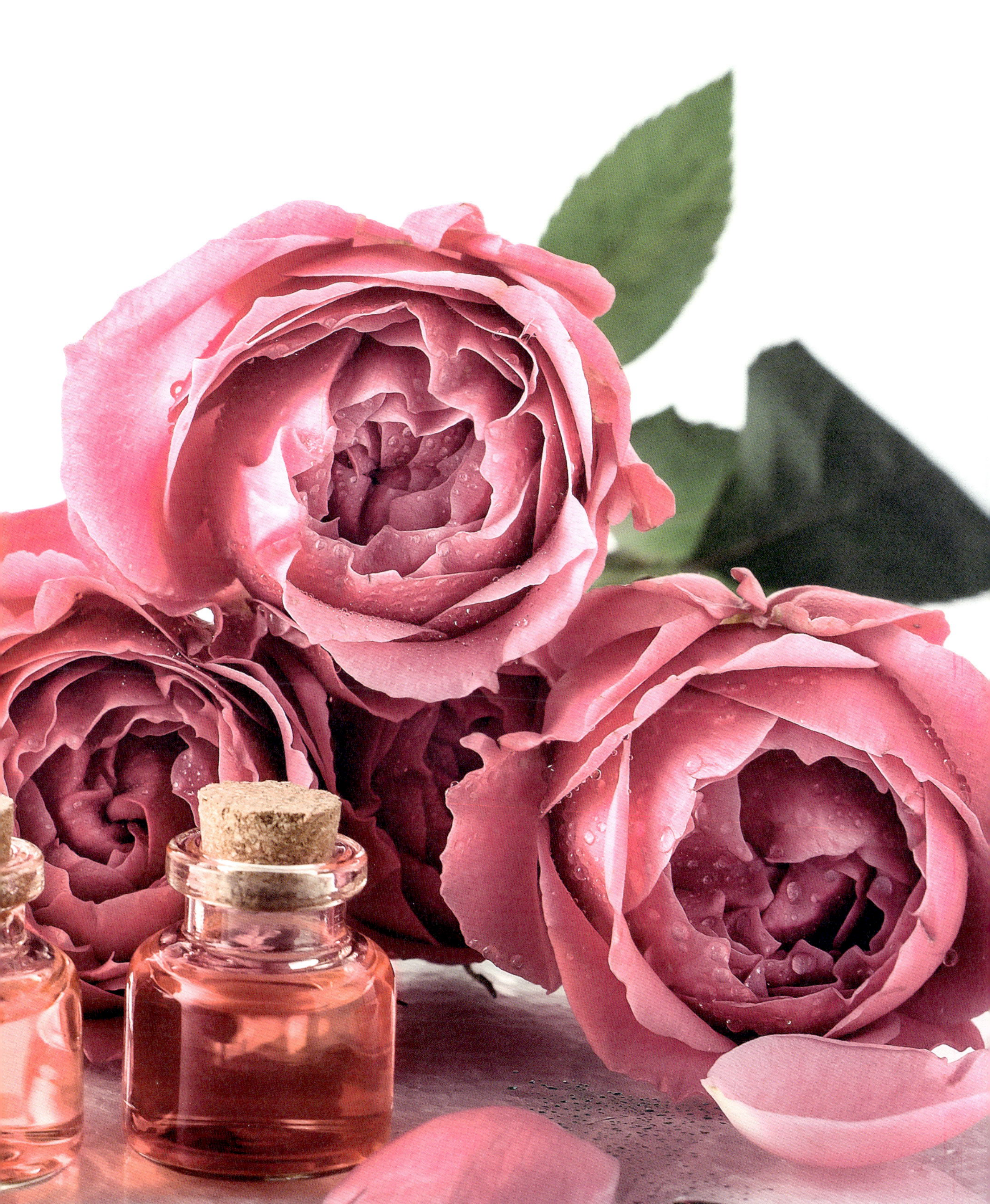

Medical use

HEALING ROSES

ACROSS CULTURES, ROSES TRADITIONALLY PLAY A MAJOR ROLE IN MEDICINE

Rose lovers know how relaxing it can be to take in the delicate fragrance of a rose. But the rose itself is nature's own *Gesamtkunstwerk*, used medicinally for centuries. Today we also know the precious compounds behind its effects: among other things, rose petals contain essential oils, tannins, and saponins whose effects are harnessed when the petals are prepared in various ways, from infusions to ointments and oils.

European herbalism and Ayurvedic medicine both make use of rose petals. Because of its astringent properties, a preparation of fresh or dried petals helps wounds heal and is said to relax and soothe inflammation. In Chinese medicine, *mei gui hua* (dried buds of *Rosa rugosa*) is traditionally used to ensure proper flow of chi or vital energy. The fruit of the rose is also packed with goodness; in the Middle Ages, long before science could establish their high vitamin content, rose hips were used to strengthen the immune system. Charlemagne even ordained that *Rosa gallica*, the apothecary's rose, was to be cultivated in all monastery gardens (p. 128).

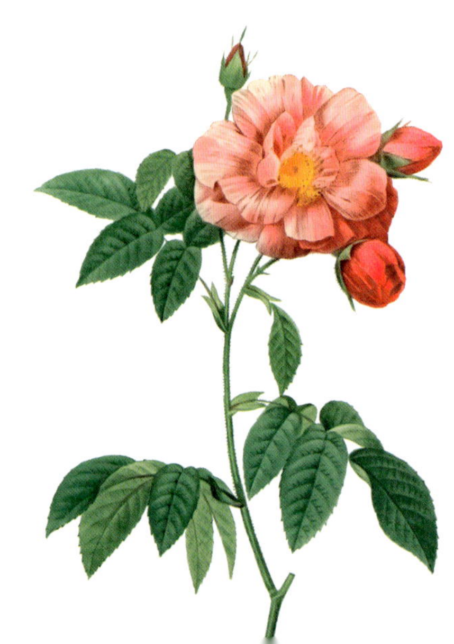

THIS PAGE | *Rosa gallica*, also known as the apothecary's rose

NEXT PAGE | In the Arab world roses were also regarded as therapeutic, as attested by this fourteenth-century Latin manuscript, itself based on an eleventh-century book created in Baghdad (left). In Chinese medicine, dried buds of *Rosa rugosa* (right) are said to maintain the flow of chi or life energy.

Medizinische Nutzung

HEILENDE ROSEN

QUER DURCH ALLE KULTUREN SPIELEN ROSEN IN DER MEDIZIN TRADITIONELL EINE GROSSE ROLLE

Rosenfreunde wissen, wie entspannend es sein kann, den zarten Duft einer Rosenblüte zu erschnuppern. Doch die Rose an sich ist ein Gesamtkunstwerk der Natur und wird schon seit Jahrhunderten medizinisch genutzt. Heute ist uns auch bekannt, welchen wertvollen Inhaltsstoffen dies zu verdanken ist: Rosenblüten enthalten unter anderem ätherische Öle, Gerbstoffe und Saponine, die in verschiedenen Zubereitungsformen vom Tee bis zu Salben und Ölen ihre Wirkung entfalten.

Sowohl die heimische Kräuterkunde wie auch die ayurvedische Heilkunde nutzen Rosenblüten: Eine Zubereitung aus frischen oder getrockneten Blütenblättern unterstützt durch die adstringierende Wirkung die Wundheilung und soll Entzündungen lindern und entkrampfend wirken. In der chinesischen Medizin nutzt man traditionell *Mei Gui Hua*, getrocknete Knospen der *Rosa rugosa*, um die Lebensenergie Qi im richtigen Fluss zu halten. Auch die Frucht der Rose hat es in sich: Hagebutten wurden schon im Mittelalter zur Stärkung des Immunsystems eingesetzt, also lange, bevor ihr hoher Vitamingehalt wissenschaftlich nachgewiesen wurde. So ist die *Rosa gallica* auch als Apothekerrose bekannt – und wurde auf Anordnung Karls des Großen in allen Klostergärten kultiviert (s. S. 132).

LINKE SEITE | *Rosa gallica*, auch als Apothekerrose bekannt

NÄCHSTE SEITE | Auch im arabischen Raum galt die Rose als Heilmittel. Links eine lateinische Handschrift aus dem 14. Jahrhundert, die auf einem in Bagdad entstandenen Buch aus dem 11. Jahrhundert basiert. Die Knospen der Kartoffelrose (*Rosa rugosa*, rechts) werden in der chinesischen Medizin getrocknet und sollen die Lebensenergie Qi im Fluss halten.

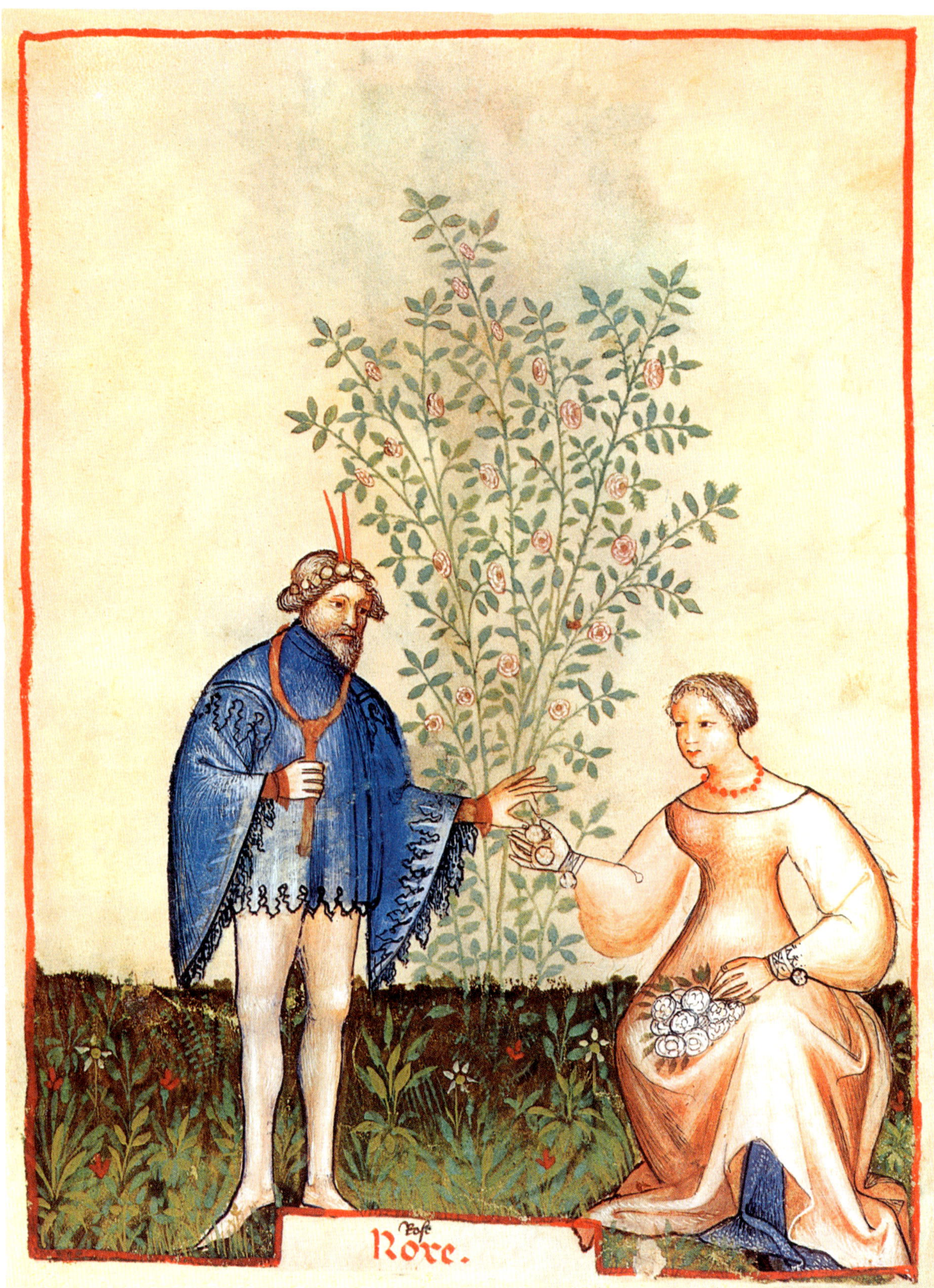
Rose
Roxe.

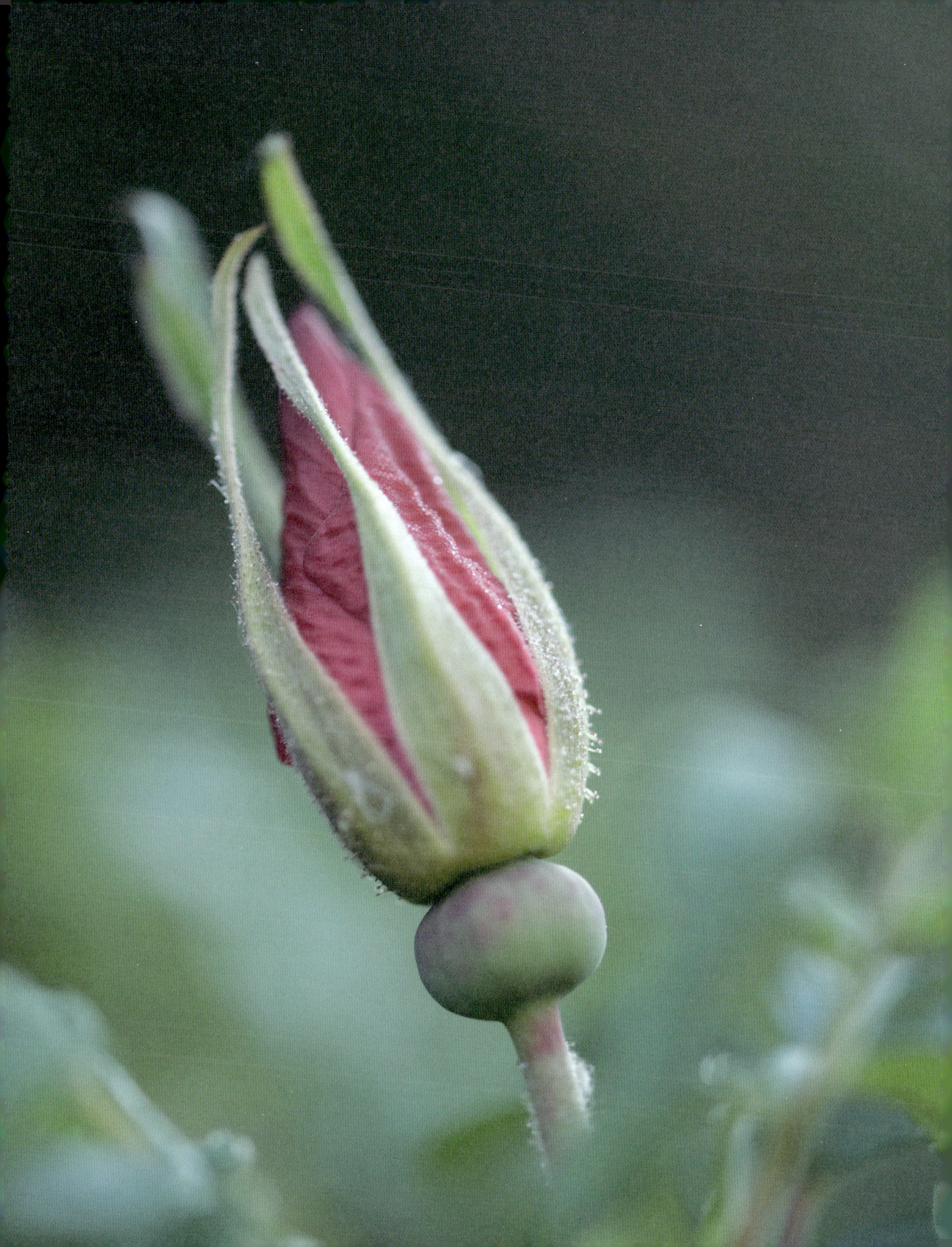

Twelfth century

HILDEGARD AND THE ROSES

SAINT HILDEGARD OF BINGEN HELPED DEFINE THE PRACTICE OF MEDICINE BY HARNESSING NATURE'S POWER TO HEAL

The abbess and naturopath Hildegard of Bingen cultivated a rich trove of medicinal knowledge during the twelfth century. She succeeded in establishing a connection between what ancient scholars had recognized on one hand and the traditional knowledge of herbalists and folk medicine on the other. From there, she worked out specific methods of healing that could be applied by anyone. Loyal adherents to some forms of alternative medicine in the German-speaking countries still refer to these methods as "Hildegard medicine".

Many of the principles Hildegard describes would later come to be recognized and proven through scientific research. She beat them to it: In her work *Physica*, Hildegard attributes cooling properties to the rose, principally to *Rosa centifolia*, and so she recommends gathering the flower petals at daybreak and laying them on irritated eyes for relief and refreshment. Hildegard also emphasizes the astringent and anti-inflammatory properties of rose flowers, which she says relieve sores and ulcers when added to salves or when applied to a skin pack. She attributes antispasmodic properties to a salve of sage leaves mixed with a double quantity of rose blossoms and boiled in fat and water. Hildegard also suggests the rose-and-sage combination for when the emotions boil over: sniffing a powder of dried roses and sage leaves was said to soothe rage and brighten the spirit.

NEXT PAGE LEFT | A statue of St. Hildegard at the monastery named after her (left). A centifolia (right), to which she ascribed cooling properties

12. Jahrhundert

HILDEGARD UND DIE ROSE

DIE KRÄFTE DER NATUR ZU HEILSAMEN ZWECKEN EINSETZEN – DAMIT WIRD HILDEGARD VON BINGEN ZUR MEDIZINISCHEN WEGWEISERIN

Die Äbtissin und Naturheilkundlerin Hildegard von Bingen trug im Laufe des 12. Jahrhunderts einen reichen Schatz an medizinischem Wissen zusammen. Ihr gelang es, die Erkenntnisse antiker Gelehrter mit überliefertem Wissen aus Kräuterkunde und Volksmedizin zu verbinden. Davon leitete sie konkrete Heilmethoden ab, die für jedermann anwendbar waren. Noch heute hat die „Hildegard-Medizin" treue Anhänger.

Viele der Wirkprinzipien, die Hildegard beschreibt, nehmen Erkenntnisse vorweg, die später wissenschaftlich untermauert wurden. Der Rose, vornehmlich der *Rosa centifolia*, schreibt Hildegard in ihrem Werk *Physica* eine kühlende Qualität zu. So empfiehlt sie, Rosenblütenblätter bei Tagesanbruch zu sammeln und sie zur Erfrischung und Beruhigung auf gereizte Augen zu legen. Auch die antientzündliche und adstringierende Wirkung von Rosenblüten unterstreicht Hildegard: Als Zusatz in Salben und Umschlägen sollen diese Linderung bei Geschwüren bringen. Entkrampfende Wirkung schreibt Hildegard einer speziellen Salbe zu: Für diese werden Salbeiblätter mit der doppelten Menge an Rosenblüten in Wasser und Fett aufgekocht. Die Rosen-Salbei-Kombination legt Hildegard auch bei überschäumenden Emotionen nahe: Ein Riechpulver aus getrockneten Rosen und Salbei soll Zorn beruhigen und das Gemüt erheitern.

NÄCHSTE SEITE | Statue der Hl. Hildegard im nach ihr benannten Kloster (links), rechts eine Zentifolie, der sie kühlende Wirkung zusprach

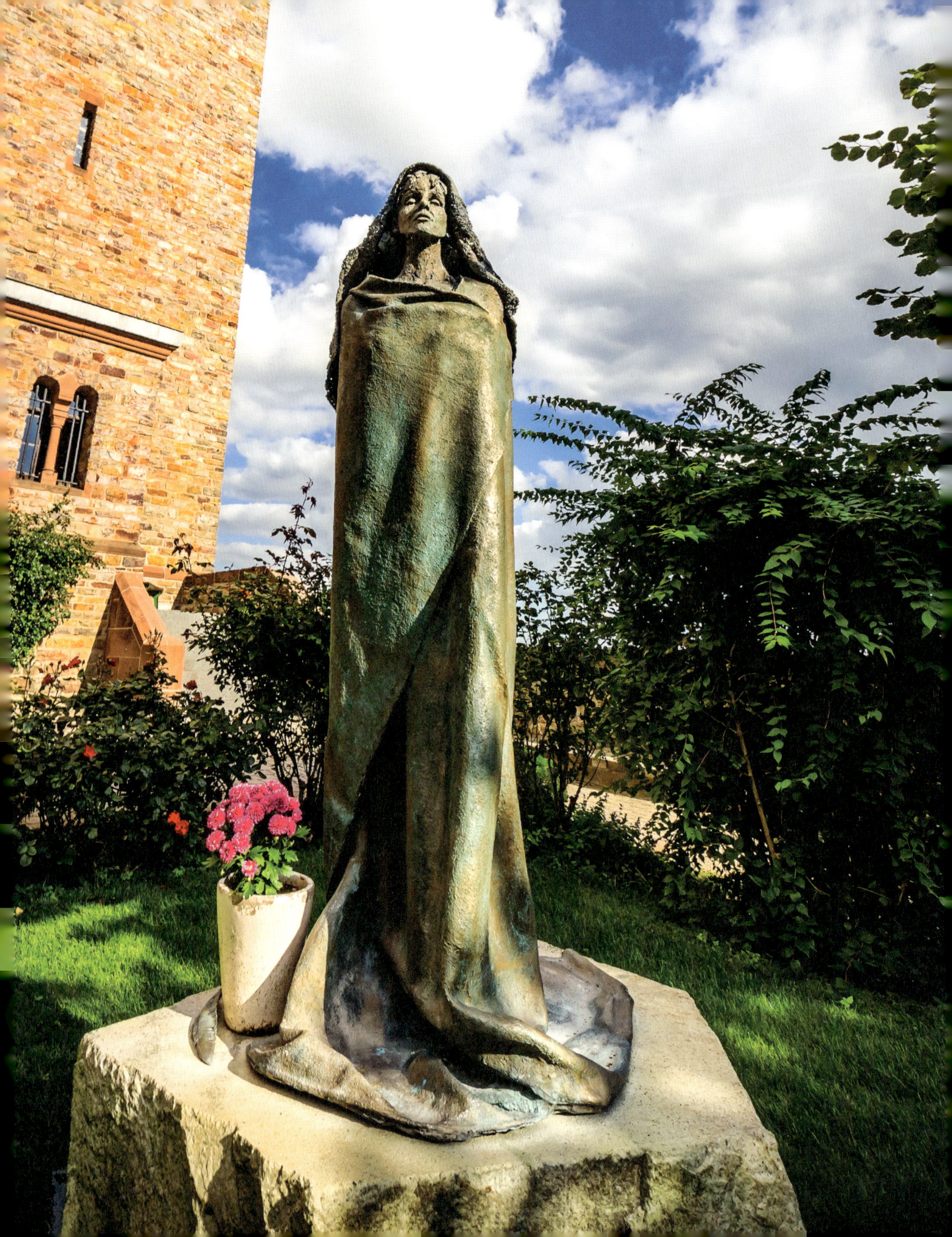

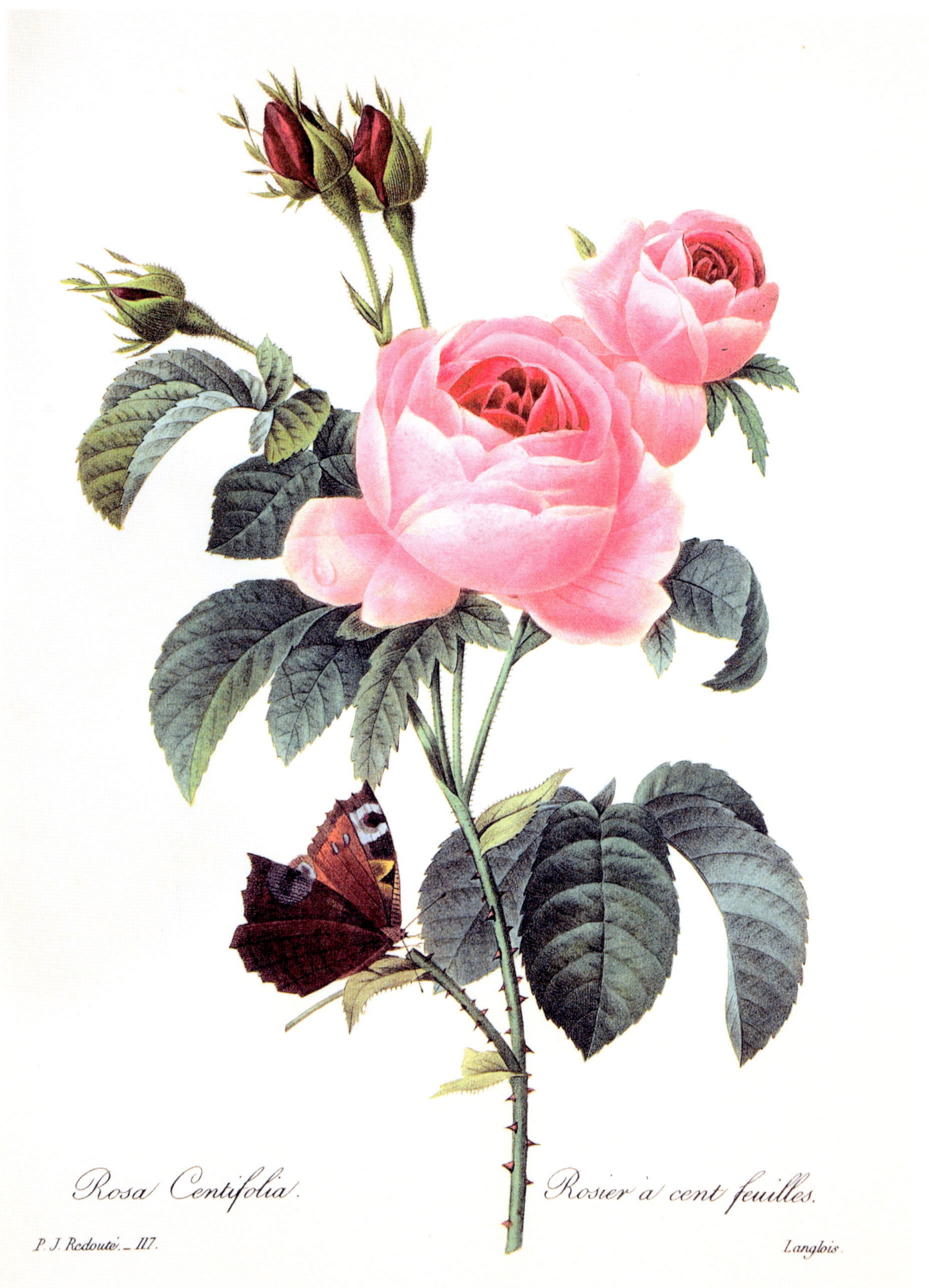

Rosa Centifolia.
Rosier à cent feuilles.
P. J. Redouté. _ 117.
Langlois.

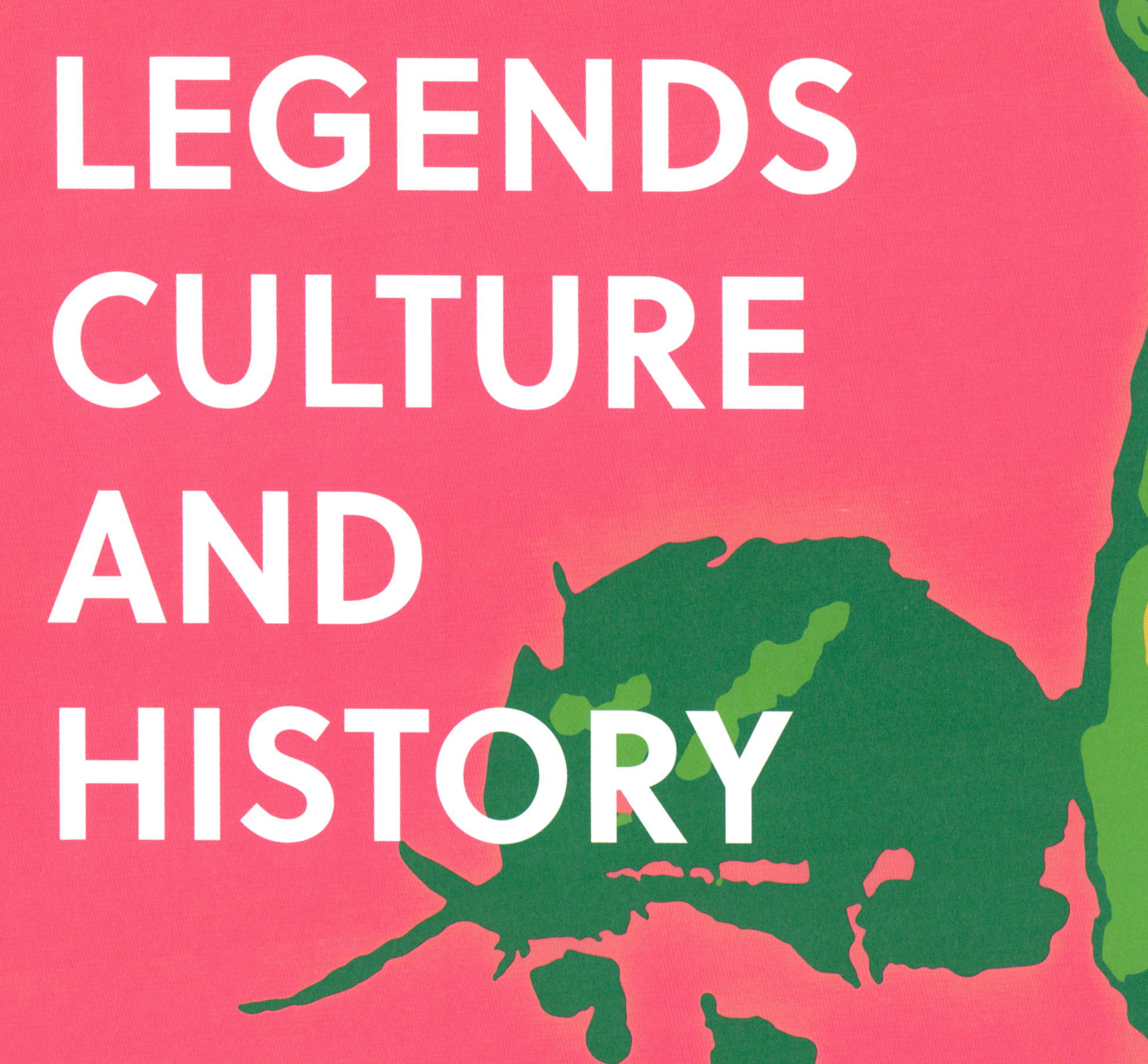

LEGENDS
CULTURE
AND
HISTORY

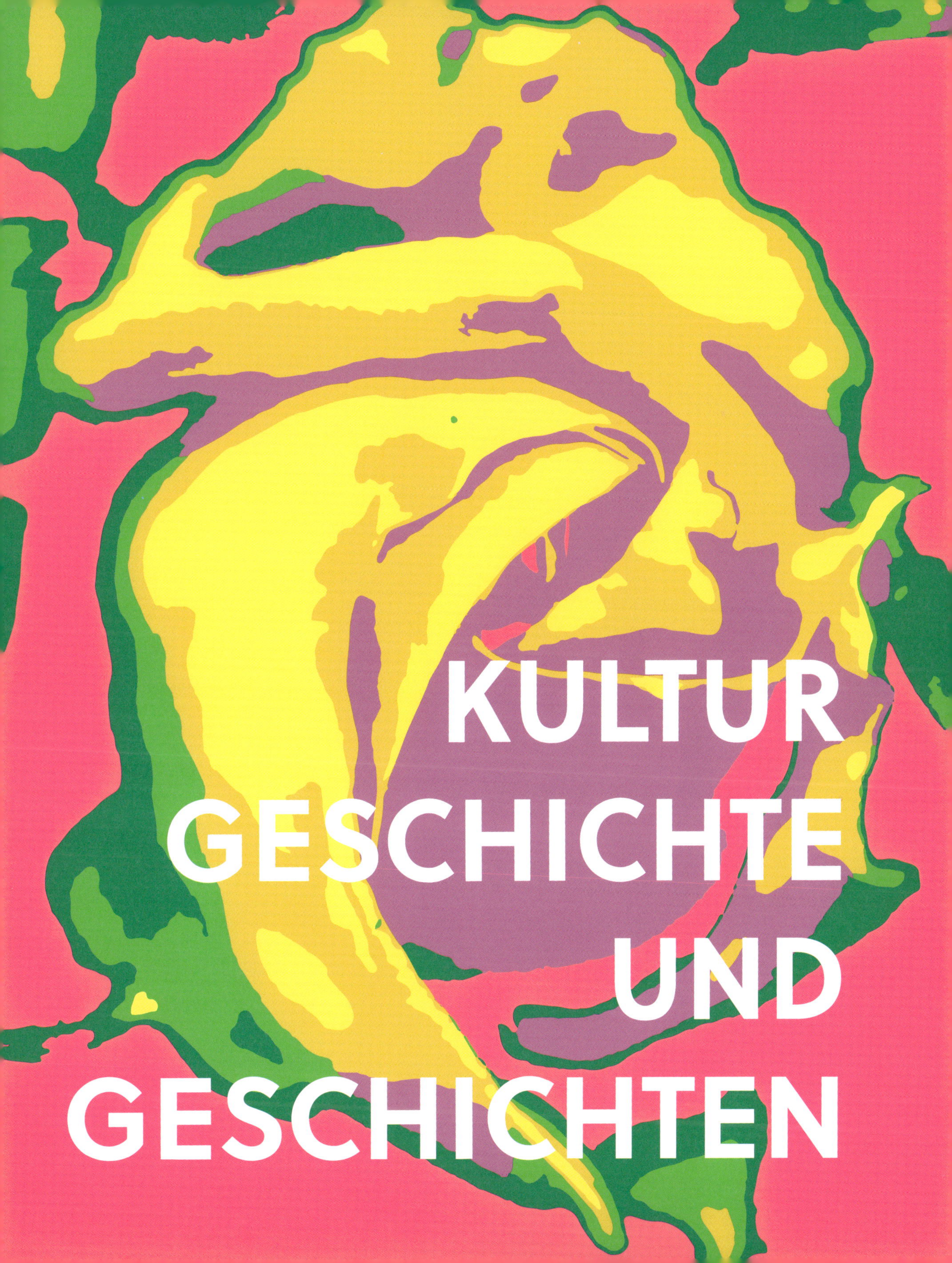
KULTUR
GESCHICHTE
UND
GESCHICHTEN

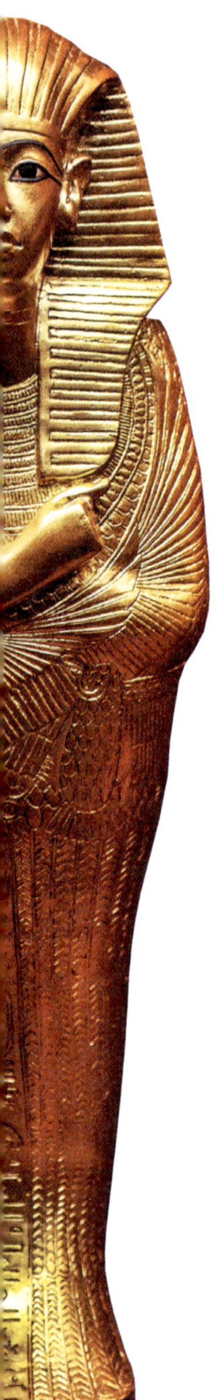

Roses in Ancient Egypt

THE DIVINE FRAGRANCE OF THE FLOWER

THE PHARAOHS BROUGHT ROSES ON THE ULTIMATE JOURNEY

To us, roses symbolize honor and devotion. In Ancient Egypt, however, this flowery symbol of adoration was oriented not primarily toward the living. Instead, flowers—particularly roses, regarded as the mark of Isis, who was worshiped as a goddess of death—were the chosen means of paying last respects.

The Ancient Egyptian cult of death is understood to have involved, as a burial rite, washing the deceased with rose water and anointing their corpses with fragrant essential flower oils. Probably, it was thought that the intense floral bouquet signaled a divine presence. Aromatic rose petals were therefore placed in tombs, as physical offerings, primarily for high-ranking individuals. And so, in 1922, when the tomb of Tutankhamun was discovered, the archaeologist Howard Carter reveled in his diary that the glint of the gold had paled in comparison to the splendor of dried flowers in the tomb. But in reality, little heed was paid to the flowery find and much more to the golden treasure. The role of aromatic flowers in the Egyptian cult of the dead was duly documented in 2006 when a plethora of petals was discovered in a sarcophagus in the Valley of the Kings, not far from Tutankhamun's tomb.

In ancient Egypt, the deceased were anointed with rose oils.
Remains of the substances have been found on coffins such
as that of Tutankhamun (here a replica).

Rosen bei den Pharaonen

DER DUFT DES GÖTTLICHEN

SCHON ÄGYPTENS HERRSCHER BEGLEITETEN ROSEN AUF IHRER LETZTEN REISE

Wir kennen Rosen als Zeichen der Hingabe und Verehrung – doch im alten Ägypten war der Liebesbeweis in Blütenform zunächst nicht vordergründig auf die Lebenden ausgerichtet: Vielmehr wurden Blumen, insbesondere auch Rosen, verwendet, um Verstorbenen die letzte Ehre zu erweisen. So gilt die Rose als Kennzeichen von Isis, die als Totengottheit verehrt wurde.

Vom altägyptischen Totenkult ist überliefert, dass als Teil des Bestattungsrituals die Verstorbenen mit Rosenwasser gewaschen und mit duftenden Blütenölen gesalbt wurden. Man glaubte wohl, intensiver Blumenduft sei ein Zeichen für göttliche Präsenz. Daher wurden aromatische Rosenblüten vornehmlich bei hochrangigen Persönlichkeiten auch als Grabbeigaben verwendet. So schwelgte der Archäologe Howard Carter im Jahr 1922 anlässlich der Entdeckung des Grabmals von Tutanchamun in seinem Tagebuch, dass die Pracht getrockneter Blüten im Grab den Glanz des Goldes regelrecht verblassen lasse – in der Realität wurde der Blütenfund dann doch zugunsten der Goldschätze vernachlässigt. Gebührend dokumentiert werden konnte die Rolle von duftenden Blüten im ägyptischen Totenkult im Jahr 2006, als man im Tal der Könige, unweit von Tutanchamuns Grab, in einem Sarkophag eine Vielzahl an getrockneten Blütenblättern fand.

Im alten Ägypten wurden Verstorbene mit Rosenölen gesalbt. An Särgen wie dem von Tutanchamun (hier eine Nachbildung) wurden Reste der Substanzen nachgewiesen.

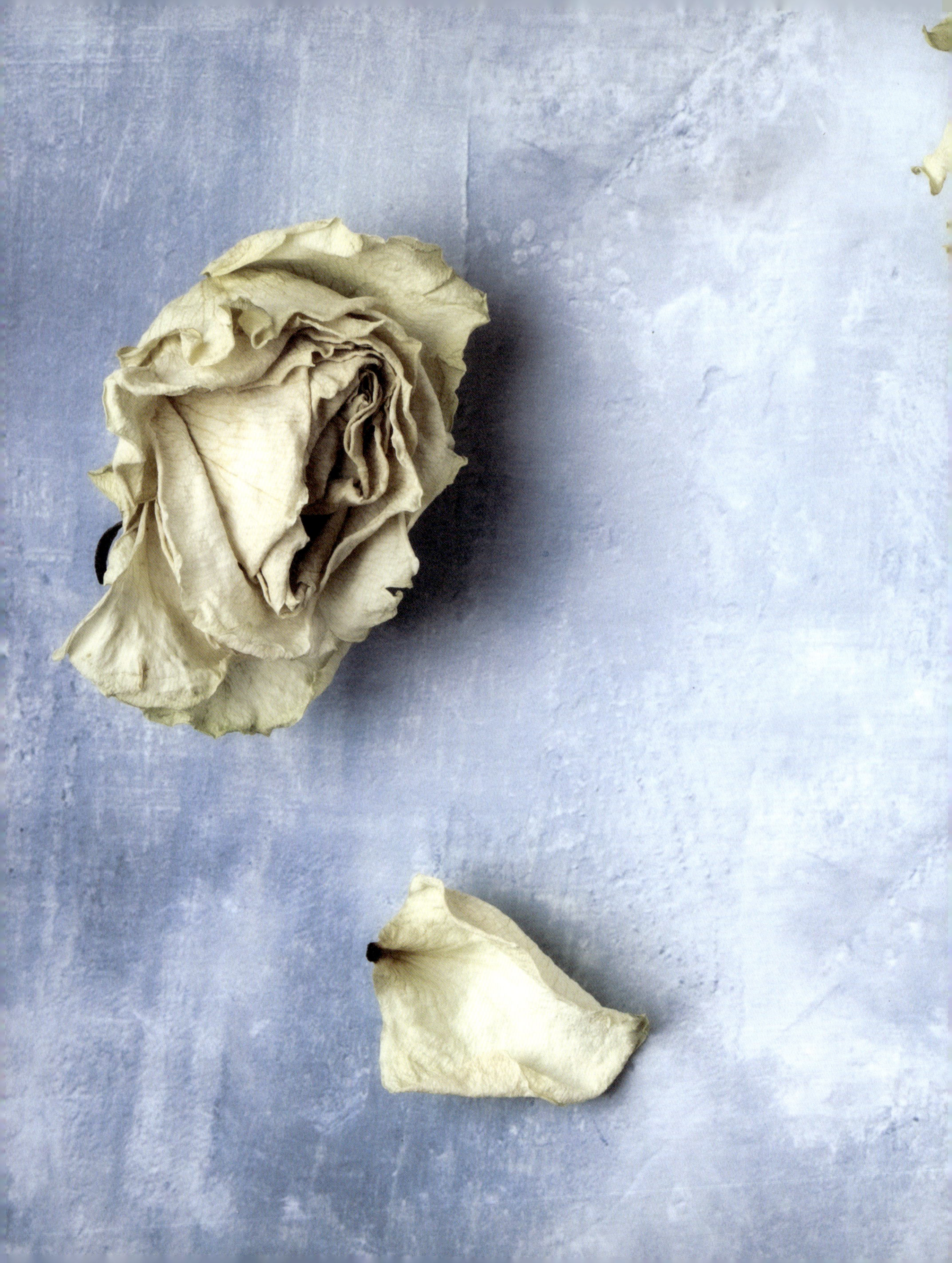

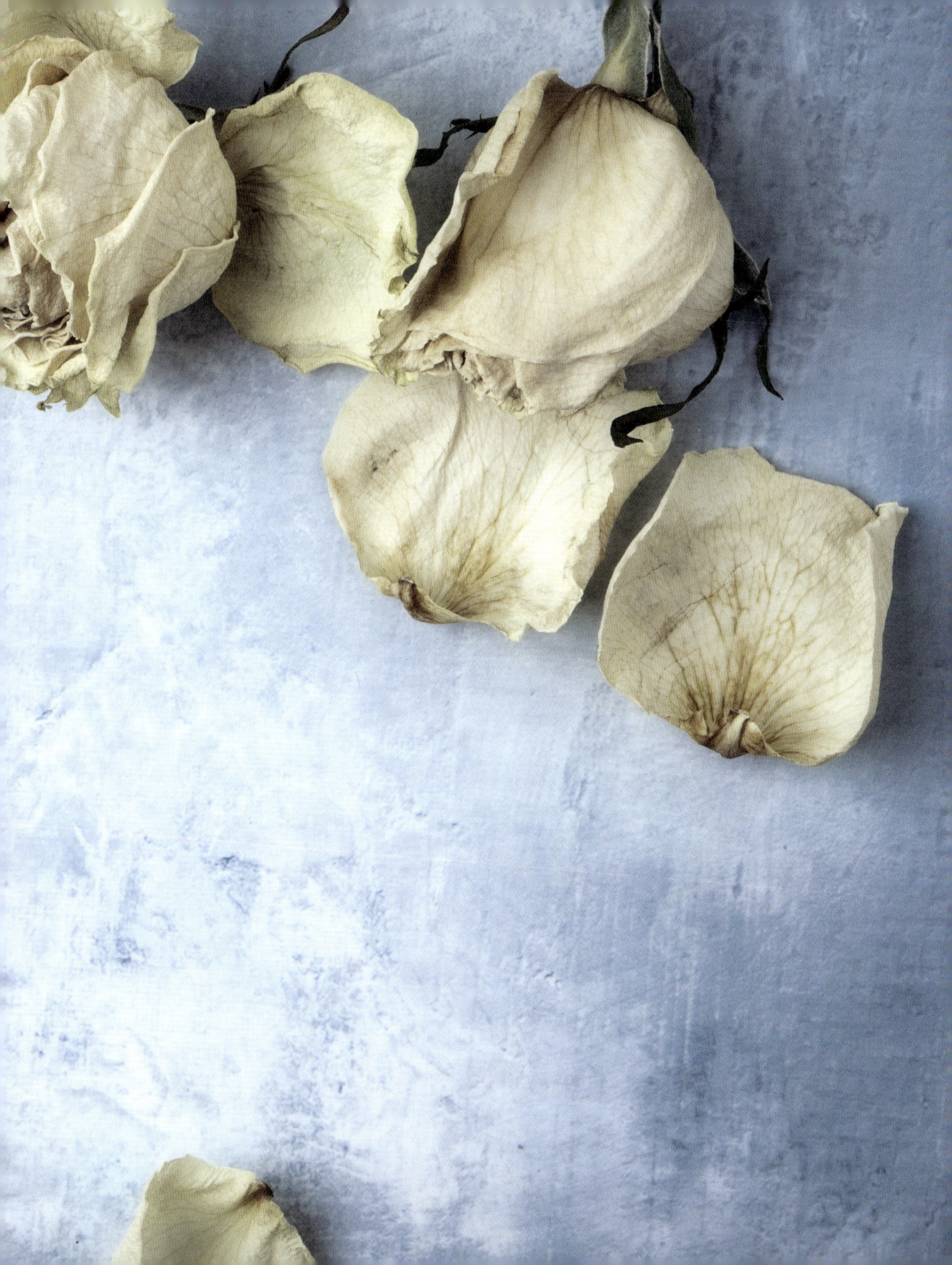

Rose cultivation in China

THE ETERNAL ROSE

IN CHINA, THE ROSE WORKS OVERTIME

Our love of roses likely extends back to ancient China, where people's appreciation for these flowering beauties has been documented for more than five thousand years. Under the Han dynasty starting in the second century BCE, so many roses were in cultivation—taking up so much arable land—that food shortages became an issue. About a thousand years ago, Chinese rose-breeders hit upon a characteristic that has made the Chinese rose very popular in the West: it blossoms again and again, and its discovery heralded the birth of the eternal rose.

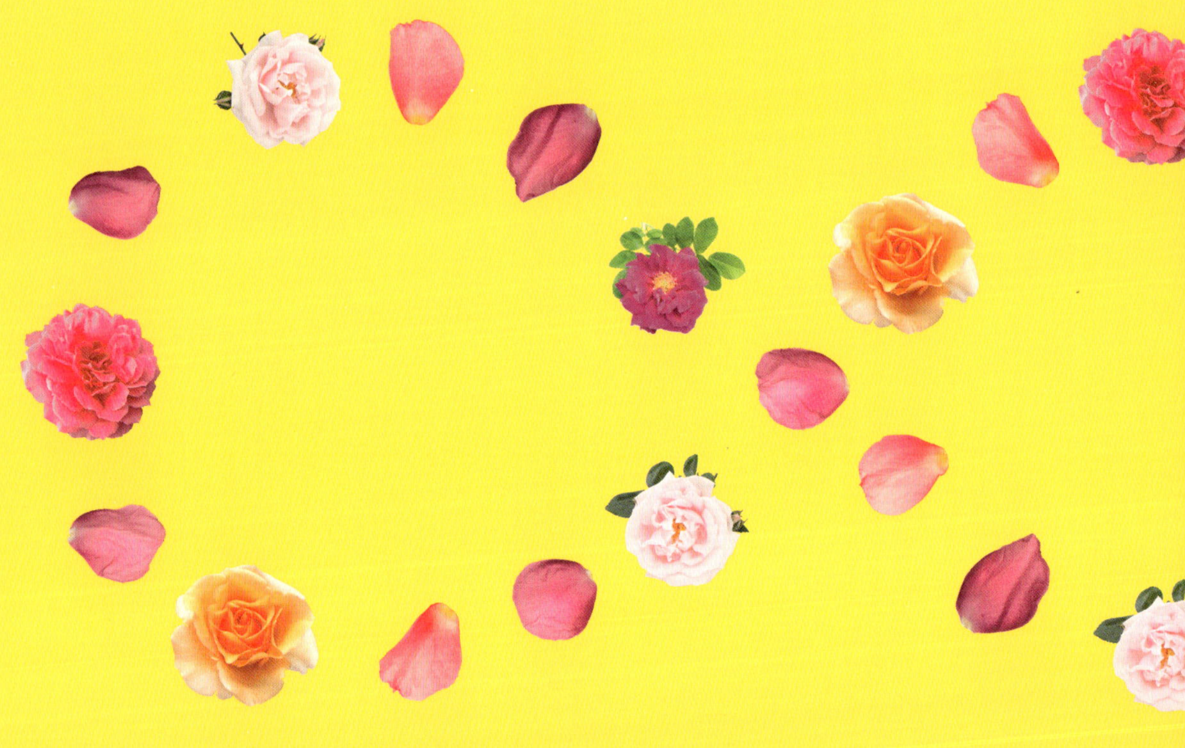

Rosenzucht in China

DIE ZEITLOSE ROSE

IM REICH DER MITTE GEHT DIE ROSENBLÜTE IN DIE VERLÄNGERUNG

Die Wurzel der Rosenliebe liegt wohl in China: Schon vor über 5000 Jahren erfreute man sich dort an den blühenden Schönheiten. In der Han-Dynastie wurden ab dem 2. Jahrhundert v. Chr. Rosen bereits in so großem Stil kultiviert, dass durch den Verbrauch fruchtbaren Landes Nahrungsengpässe drohten. Vor rund 1000 Jahren entdecken Rosenzüchter in China ein Charakteristikum, das die China-Rose auch bei uns so beliebt macht: Sie kam übers Jahr wieder und wieder zum Blühen – die zeitlose Rose war geboren.

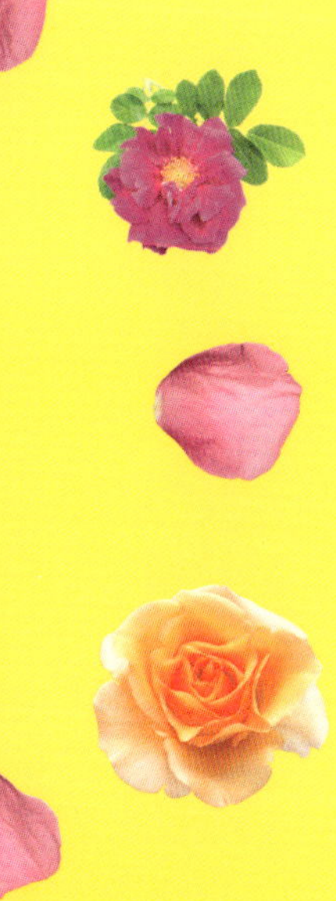

Commemoration of the dead

ROSALIA

ROMAN FESTIVAL OF ROSES

Each year in May, in the temperate latitudes, the blossom begins. So it is no surprise that this season—from May until the principal blossom in early summer—was when the ancient Romans traditionally celebrated Rosalia. Sometimes called the *rosaria* or *dies rosationis* (the day of the rose), Rosalia grew out of the many ways of commemorating the dead. On certain holidays, it was customary to honor deceased family members by laying flowers on their tombs, first violets (the *dies violationis* marking the occasion) and on through the festivals of roses from May to July. The cult of roses was so popular that enormous fields of roses were cultivated outside the gates of Rome. It would obviously have been important to procure an ample supply of rose petals for Rosalia.

Another kind of religious cult of roses has been resurrected in Rome in more recent times, by the way, and it is spectacular: on the feast of the Pentecost, at the height of the rose blossom, red roses rain down for minutes at a time from the circular "eye" in the roof of the Pantheon. Today, they symbolize the flame of the Holy Spirit.

NEXT PAGE | At Pentecost it ‚rains' rose petals in the Pantheon in Rome.

Totengedenken in Rom

DAS ROSALIA-FEST

DEN VORFAHREN ZU EHREN WURDEN GRÄBER MIT ROSEN GESCHMÜCKT

Jedes Jahr im Mai beginnt in unseren Breitengraden die Zeit der Rosenblüten. So ist es auch kaum verwunderlich, dass eben in diesem Zeitraum, ab Mai bis zur Hauptblütezeit im frühen Sommer, schon im alten Rom traditionell das Rosalia-Fest zelebriert wurde. Bisweilen auch als *rosaria* oder *dies rosationis* (Tag des Rosenschmucks) bezeichnet, entstand das Rosenfest aus den zahlreichen Arten des Totengedenkens: Um die verstorbenen Familienmitglieder zu ehren, war es Brauch, zu bestimmten Feiertagen Blumen auf deren Grabmälern niederzulegen – beginnend mit den ersten Veilchen anlässlich der *dies violationis* bis zu den Rosenfesten, die zwischen Mai und Juli stattfanden. Derart beliebt war der Rosenkult, dass man vor den Toren Roms sogar speziell dafür riesige Rosenfelder anlegte. Schließlich wollte man sichergehen, dass zu Rosalia auch genügend duftende Rosenblüten zur Verfügung stehen würden!

In der Neuzeit wurde übrigens in Rom auch eine andere Art des religiösen Rosenkults wieder zum Leben erweckt: Es ist schon ein eindrucksvolles Schauspiel, wenn es zum Pfingstfest, mitten in der Zeit der Rosenblüte, aus dem kreisrunden „Auge" im Dach des Pantheons minutenlang rote Rosenblüten regnet – nun allerdings als Symbol für die Flammen des Heiligen Geistes.

NÄCHSTE SEITE | An Pfingsten „regnet" es Rosenblüttenblätter im Pantheon in Rom.

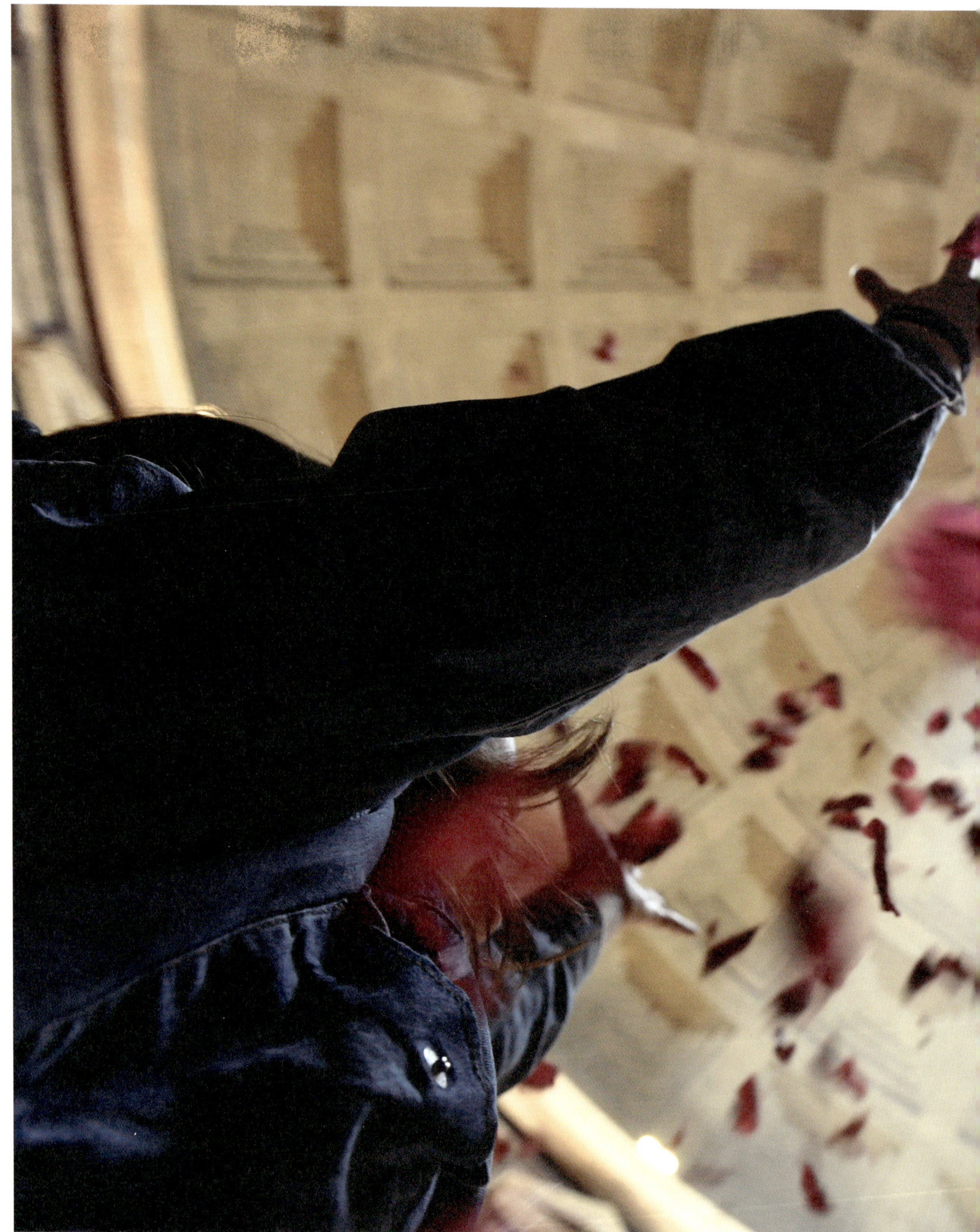

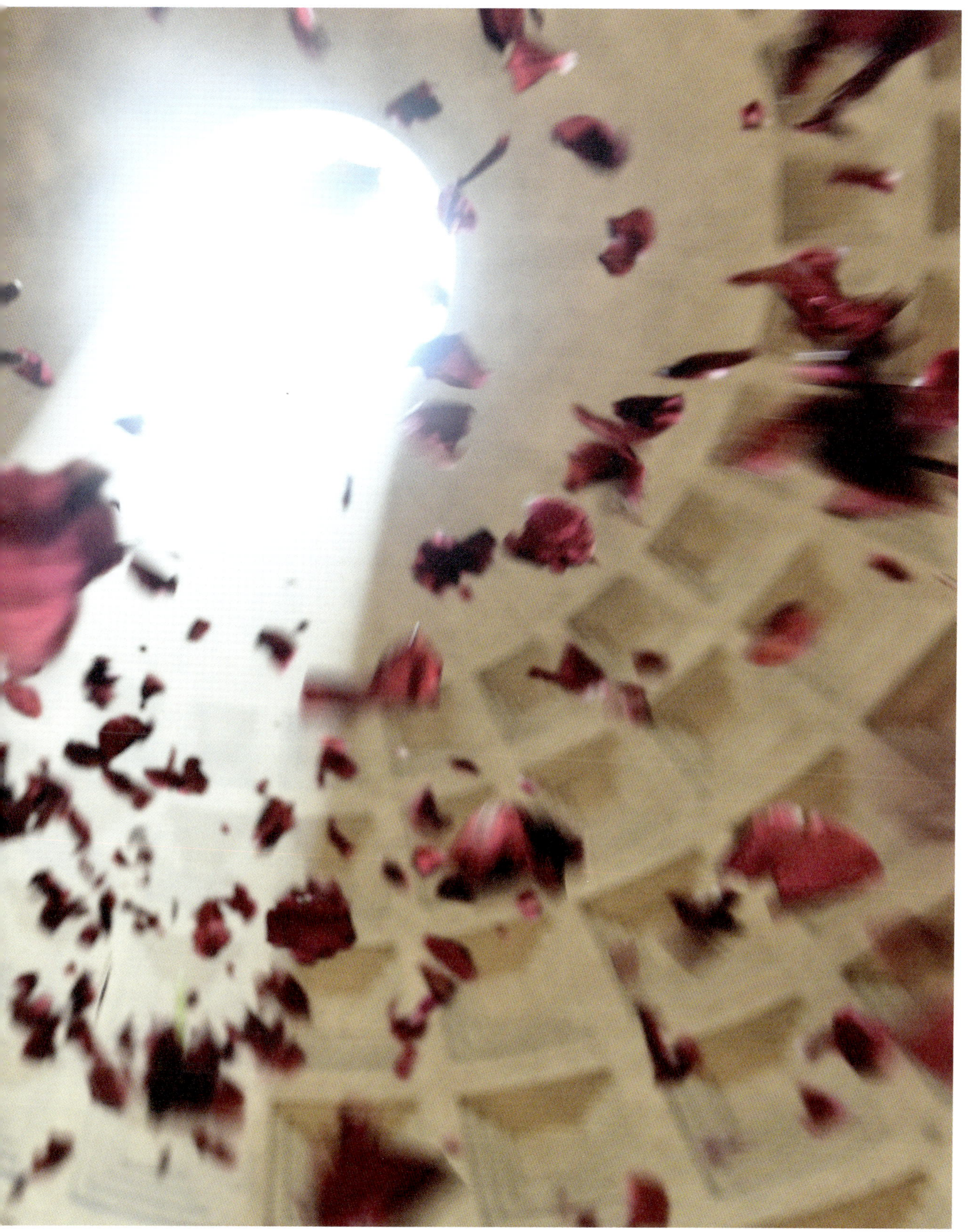

Roses in the ancient world

LOVE AND LUXURY

APHRODITE, VENUS, AND CLEOPATRA:
ROSES FOR LOVE

We have the Greek poet Sappho to thank for the rose having become known as the "queen of flowers" in antiquity, raving as she did, as early as 600 BCE, that "the rose the queen of flowers should be." The first known pictorial depiction of the rose traces back even further: depicted in a fresco (from around 1600 BCE, according to estimates) from the Minoan palace of Knossos on the Greek island of Crete, are delicate, five-petaled roses.

In Greek mythology, the rose became a symbol of love and beauty as an attribute of the goddess Aphrodite. According to one ancient legend, Aphrodite fell in love with Adonis, but Adonis's other passion was hunting. Disquieted by ominous forebodings, Aphrodite warns Adonis to be wary of wild animals. Adonis brushed off her admonitions—a fatal error, as it would turn out. Adonis, on the prowl, paid with his life when he encountered Ares, the god of war, who, jealous of his rival, had assumed the guise of a wild boar. Aphrodite wept bitterly for her beloved, and where her tears mixed with his blood, there grew up a rose bush, a sign of their everlasting love.

In Roman mythology, too, there is a legend about Cupid, son of Venus the goddess of love. The rose's origin is attributed to him. But considering that he is love's messenger, the tale is not as romantic as you might think: While Cupid was carrying nectar to the gods, he spilled a few drops of the heavenly drink, and where these fell to the ground, roses ever would grow.

Aside from tales and legends it is known that for Romans, roses were a sought-after luxury. Fragrant rose petals were used in perfumery and to flavor wines. Of the debauched feasts for which the emperor Nero was

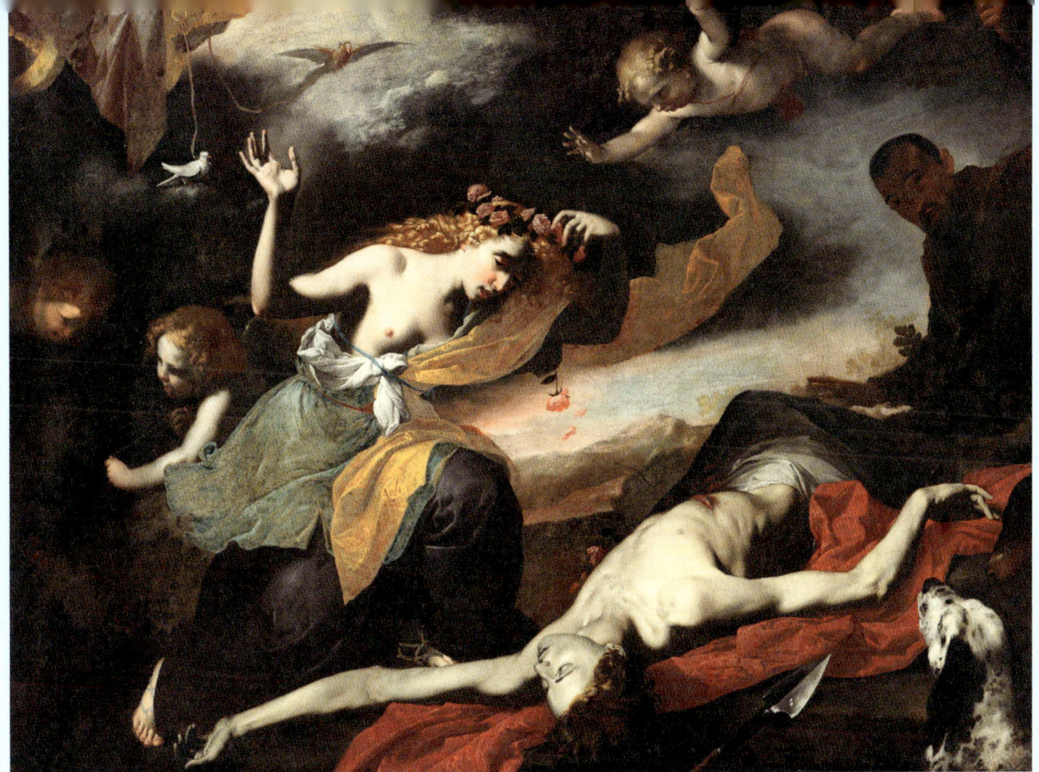

Venus Discovering the Dead Adonis, anonymous
painter from Naples, c. 1650

Venus entdeckt den toten Adonis, unbekannter
Maler aus Neapel, ca. 1650.

infamous, it was reported that the guests bathed in rose water while
rose petals fluttered and rose oil sprinkled down from the ceiling.

Cleopatra, too, is said to have had her chambers so opulently decorated before a night of lovemaking with Marcus Antonius that he had to
wade his way through knee-deep rose blossoms to reach her. So it is no
wonder that roses gradually took on a reek of decadence and profligate
obsession.

NEXT PAGE | A detail of Botticelli's painting
The Birth of Venus. The nymph Chloris can be
seen becoming the goddess of spring blossom
in the embrace of the west wind Zephyr. She
exhales rose petals when she speaks.

NÄCHSTE SEITE | Detail von Botticellis Gemälde
Die Geburt der Venus. Zu sehen ist die Nymphe
Chloris, die in der Umarmung des Westwinds Zephyr
zur Göttin der Frühlingsblüte wird. Und wie erkennt
man sie? Sie atmet beim Sprechen Rosenblüten aus.

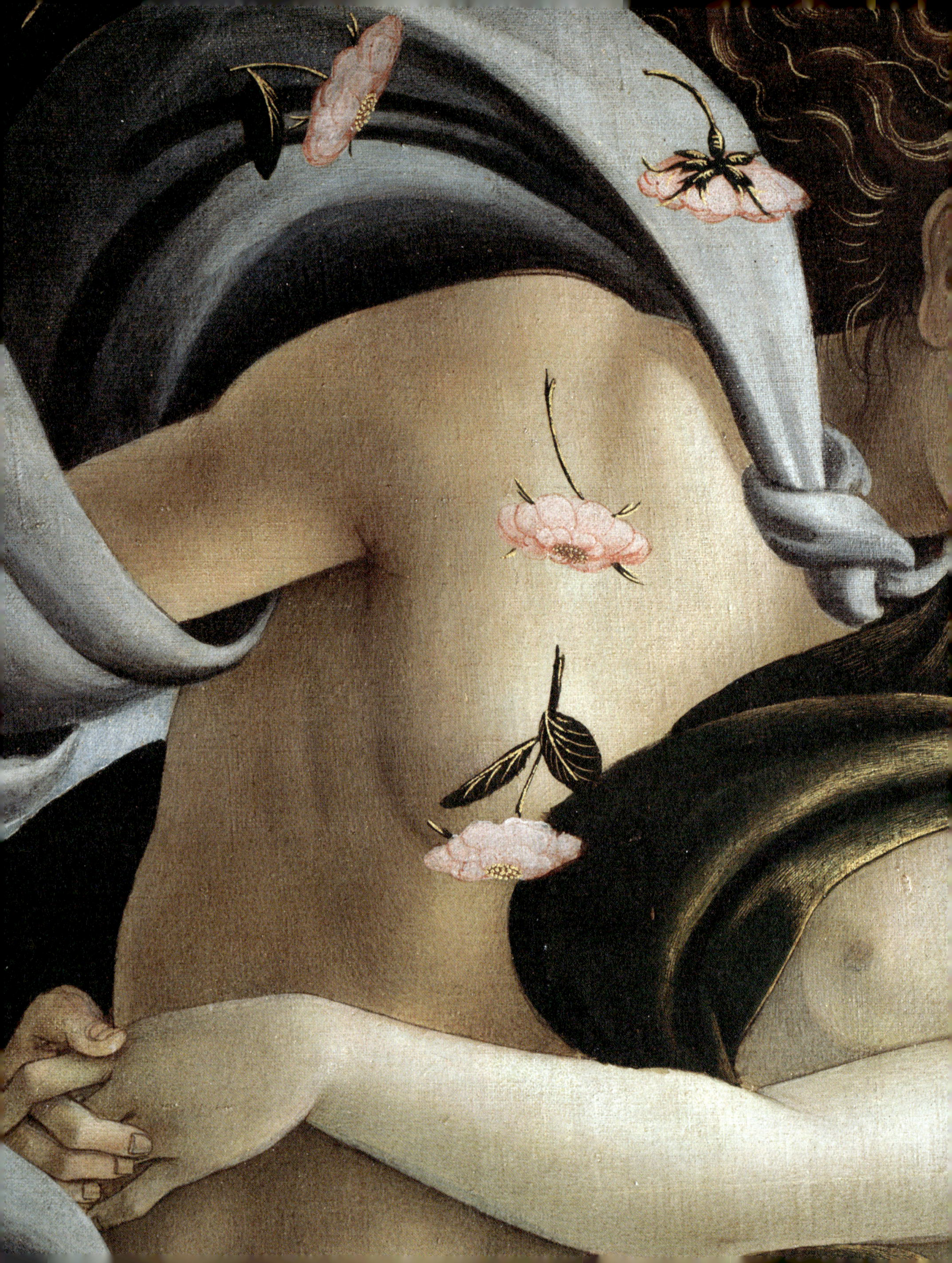

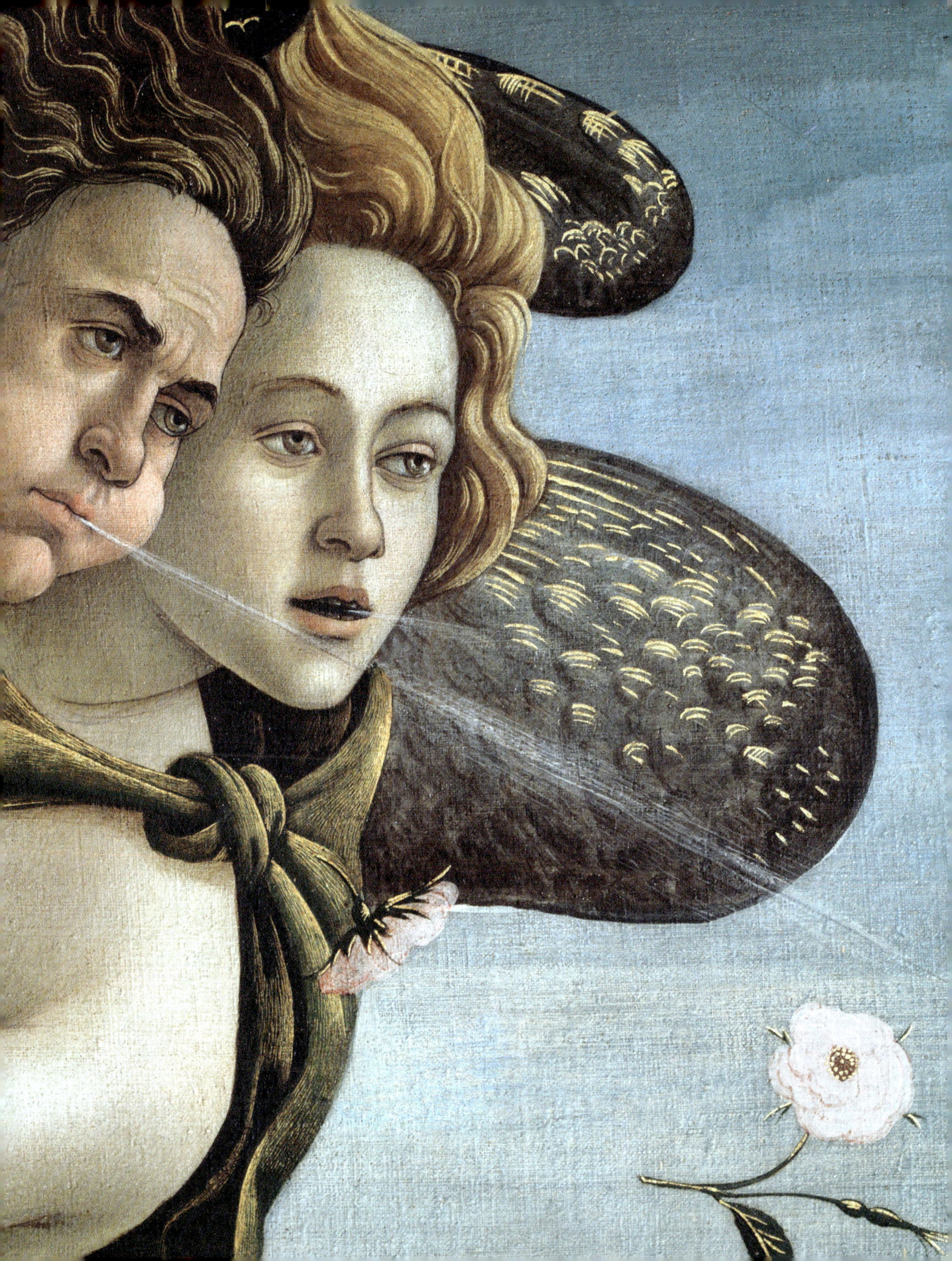

Rosen in der Antike

LIEBE UND LUXUS

APHRODITE, VENUS UND KLEOPATRA – ROSEN FÜR DIE LIEBE

Dass die Rose schon in der Antike als „Königin der Blumen" bekannt wurde, haben wir der griechischen Dichterin Sappho zu verdanken, die bereits um das Jahr 600 v. Chr. schwärmte: „Oh die Rose, ach, die Rose ist der Blumen Königin!" Noch deutlich weiter zurückverfolgen lässt sich die erste uns bekannte Rosendarstellung in bildlicher Form – zarte fünfblättrige Rosenblüten sind zu erkennen auf einem Fresko aus dem minoischen Palast von Knossos auf der griechischen Insel Kreta, dessen Ursprung auf ca. 1600 v. Chr. geschätzt wird.

In der griechischen Mythologie wurde die Rose als Attribut der Göttin Aphrodite zum Symbol für Liebe und Schönheit. Eine der alten Legenden erzählt, dass Aphrodite sich in Adonis verliebte – doch dessen Leidenschaft gehörte auch der Jagd. Aphrodite, von unheilvollen Vorahnungen geplagt, warnte Adonis, vor wilden Tieren auf der Hut zu sein. Adonis aber schlug ihre Warnungen in den Wind – fatal, wie sich herausstellen sollte. Offenbar hatte sich Kriegsgott Ares aus Eifersucht über seinen Rivalen in einen wilden Eber verwandelt. Als Adonis diesem auf der Pirsch begegnete, bezahlte er mit seinem Leben. Aphrodite beweinte ihren Geliebten bitterlich, ihre Tränen vermischten sich mit seinem Blut – und daraus erwuchs als Zeichen ihrer immerwährenden Liebe ein Rosenstrauch.

Auch in der römischen Sagenwelt gibt es eine Legende um Cupido, den Sohn der Liebesgöttin Venus, dem die Entstehung der Rose zugeschrieben wird. Diese Geschichte ist allerdings weniger romantisch, als man es bei einem Liebesboten vermuten würde: Als Cupido den Göttern ihren Nektar bringen wollte, verschüttete er ein paar Tropfen des himmlischen Göttertranks – und dort, wo diese auf die Erde fielen, wuchs fortan ein Rosenstrauch.

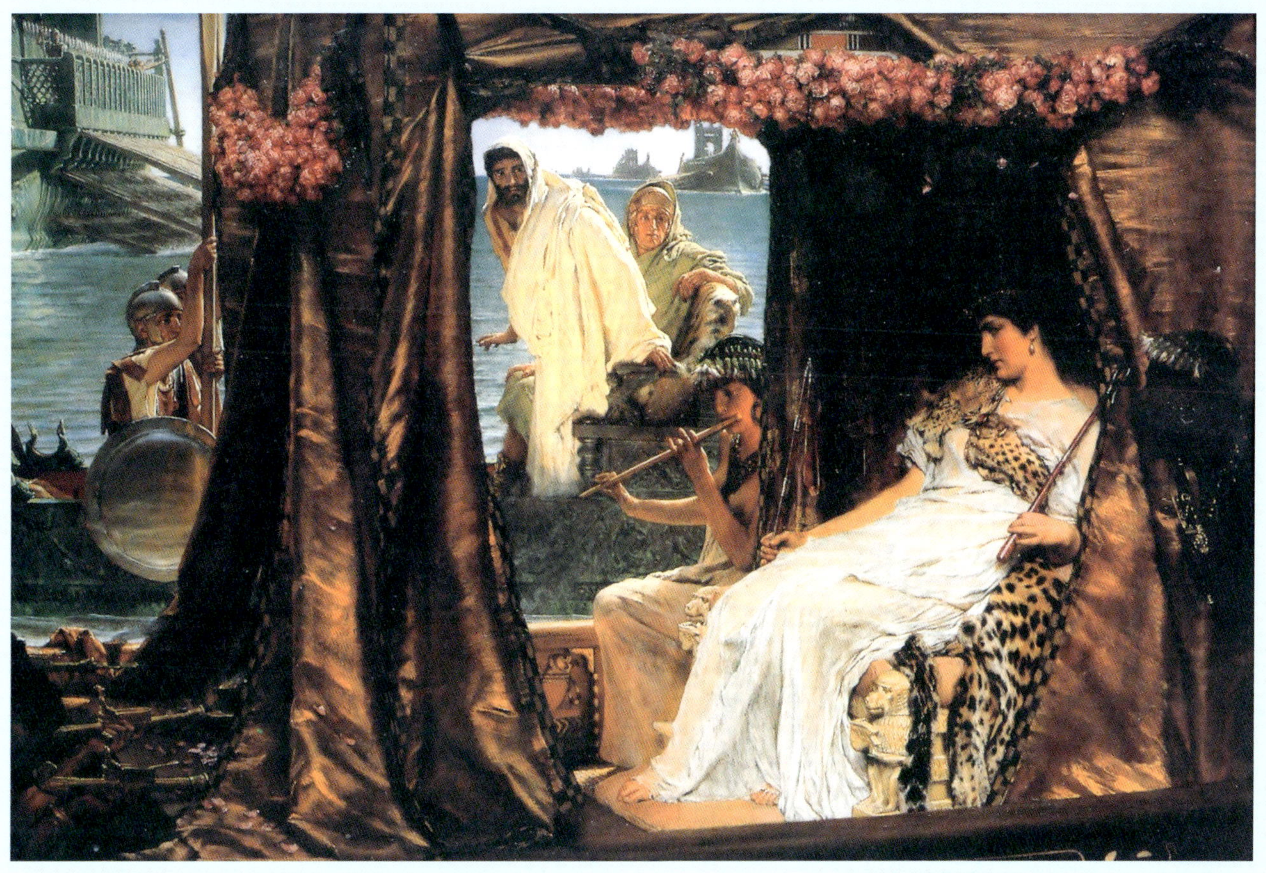

Lawrence Alma-Tadema: *The Meeting of Antony and Cleopatra* (1885)

Lawrence Alma-Tadema: *Kleopatra empfängt Marcus Antonius* (1885)

Jenseits von Sagen und Legenden ist aber überliefert, dass Rosen für die Römer ein begehrtes Luxusgut waren: Man stellte daraus Parfüm her und aromatisierte Wein mit duftenden Rosenblüten. So wird berichtet, dass bei den ausschweifenden Festgelagen, für die auch Kaiser Nero berühmt war, die Gäste in Rosenwasser badeten, während von der Decke Rosenblüten und Rosenöl auf sie herabrieselten. Auch von Kleopatra ist eine ähnlich schwelgerische Vorliebe für Rosen überliefert: Angeblich ließ sie vor einer Liebesnacht mit ihrem Geliebten Marcus Antonius ihre Gemächer so opulent schmücken, dass dieser sich seinen Weg zu Kleopatra knietief durch Rosenblüten bahnen musste. Kein Wunder also, dass die Rose sich so allmählich einen Ruf von Dekadenz und sündiger Verschwendungssucht eroberte.

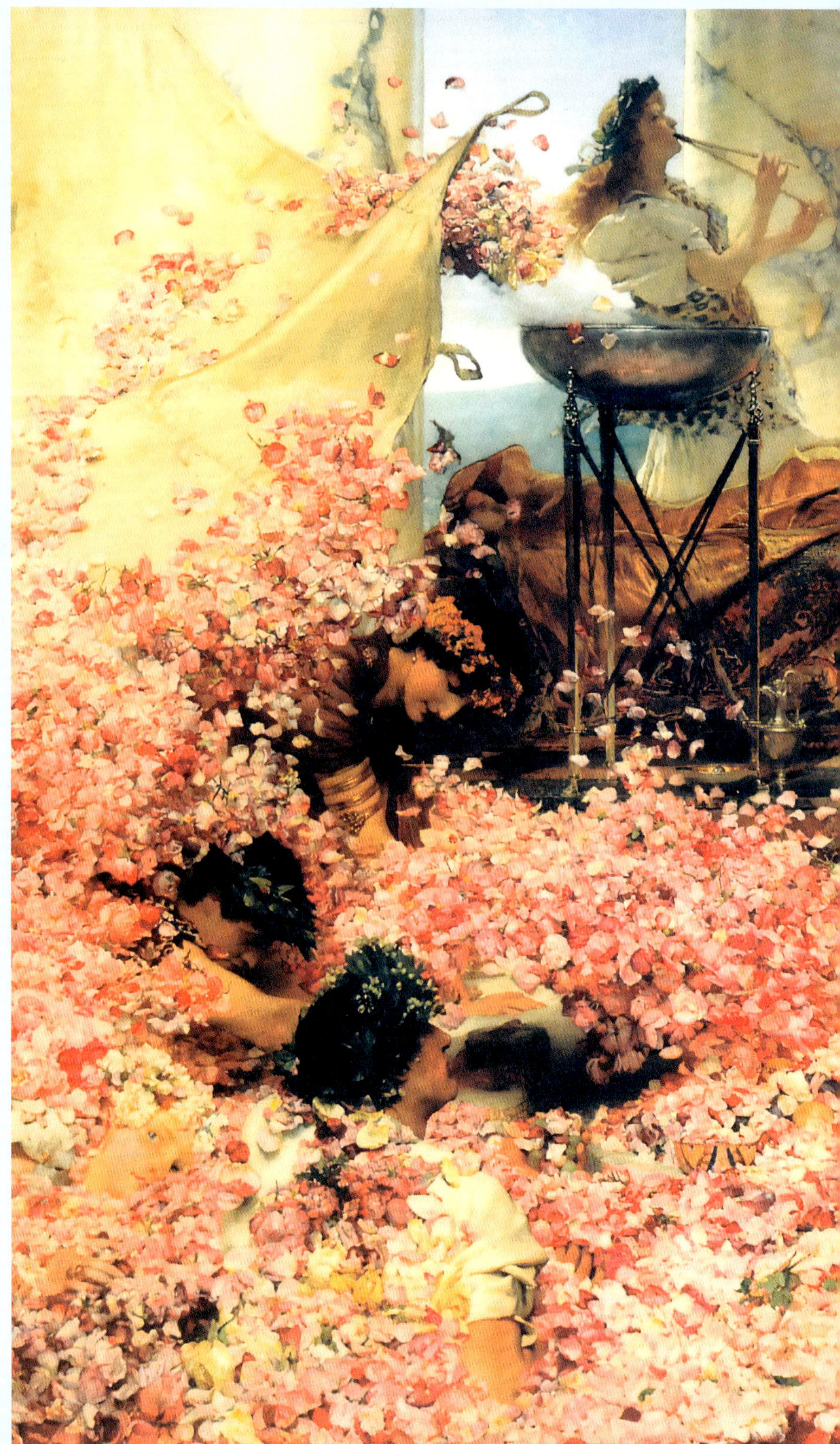

A lavish feast depicted by the Dutch
painter Lawrence Alma-Tadema
(1836–1912): In the palace of the
youthful third-century Roman emperor
Elagabalus, his guests are all but
engulfed in rose petals; some are even
said to have suffocated.

Ein rauschendes Fest in der Darstellung
des niederländischen Malers Lawrence
Alma-Tadema (1836–1912): Im Palast des
jugendlichen römischen Kaisers
Elagabal (3. Jh. n. Chr.) versinken seine
Gäste geradezu in Rösenblättern,
einige sollen sogar erstickt sein.

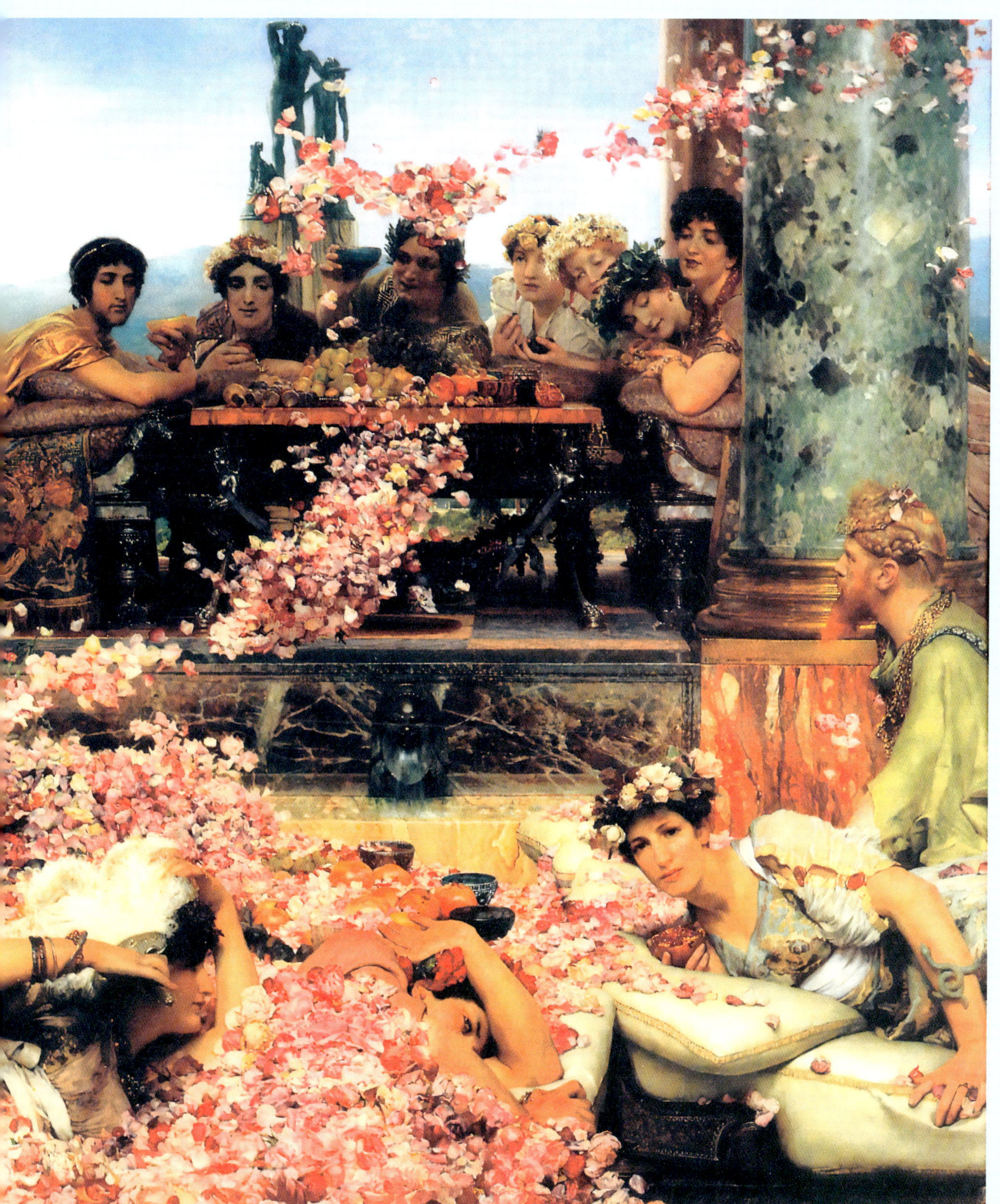

Welcome to the Middle Ages!

THE TRANSFORMATION OF THE ROSE

A SYMBOL OF DECADENT PLEASURE BECOMES THE EPITOME OF VIRGIN PURITY

Different times, different mores: between Classical Antiquity and the Middle Ages, the rose as a symbol of love underwent a marked paradigm shift. In ancient Rome, the idea of roses was one of intoxicating aroma and the flower's sensuous glory that recalled the image of debauched parties and sumptuous opulence. Early Christianity came and adopted a proven and consistently successful tactic: Instead of banning pagan symbols and customs, they were simply hijacked and reinterpreted in line with Christian values. For example, the Romans' Saturnalia became Christmas, the feast of the birth of Christ. The rose experienced a similar shift as it was reinterpreted from a symbol of sinful doings into one of the most adored symbols in Christianity. As such it remains immensely popular. This legitimized the cultivation of roses in many monastery gardens again, fortunately restoring the therapeutic flower to its beneficial duties, this time with the blessing of the Church.

In the Christian Middle Ages, a white rose was elevated to the emblem of Mary, a sign of virgin purity and chaste asceticism. To this day, numerous church hymns laud the Virgin Mary as the "rose without a thorn". And red roses on the other hand went from being a symbol of sin and lust to signifying passion and devotion to one's love for God. The depiction of rose vines with wild thorns is considered a symbol of martyrdom and of hoping for redemption—and so Jesus's crown of thorns is sometimes adorned with red roses.

Medieval depictions of roses show a strict geometry of five evenly spaced petals, another indication that the pentagram was quite deliberately incorporated into Christian nomenclature from pagan magic cults. One of the oldest floral motifs in art and architecture is a stylized rose in many medieval churches and cathedrals, from the rosette as the keystone of a groin vault to magnificent stained-glass rose windows. This is probably how the rose came to be a symbol on coats of arms.

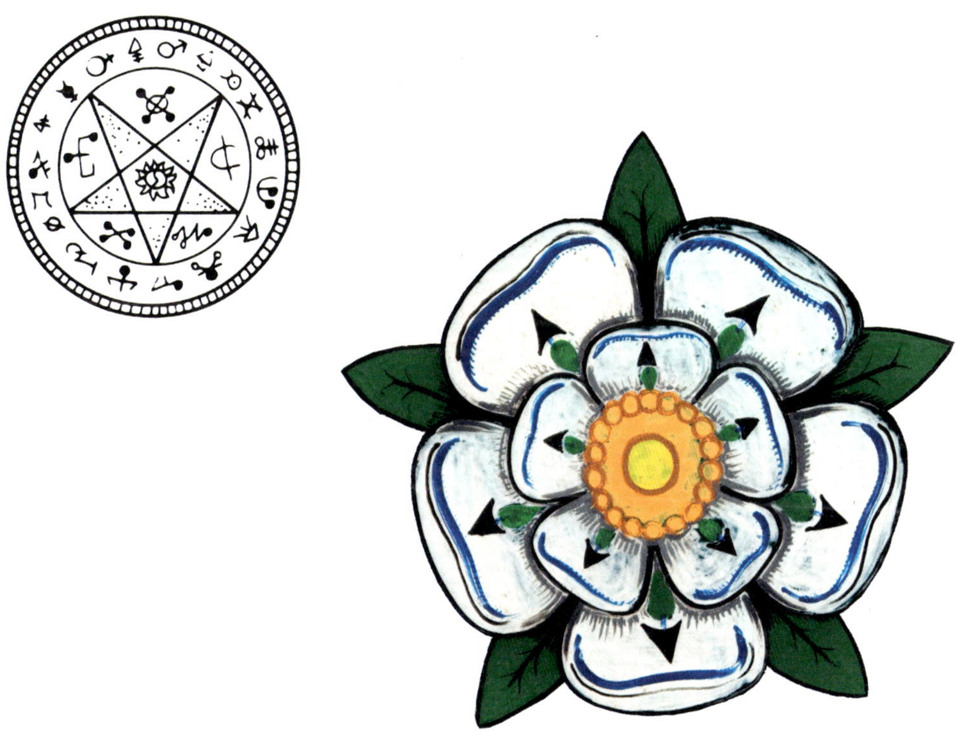

Stylised rose of York with five petals –
a telltale resemblance to the pentagram
from pre-Christian times

Stilisierte Rose von York mit fünf
Blütenblättern – dem Pentagramm aus
vorchristlicher Zeit verräterisch ähnlich

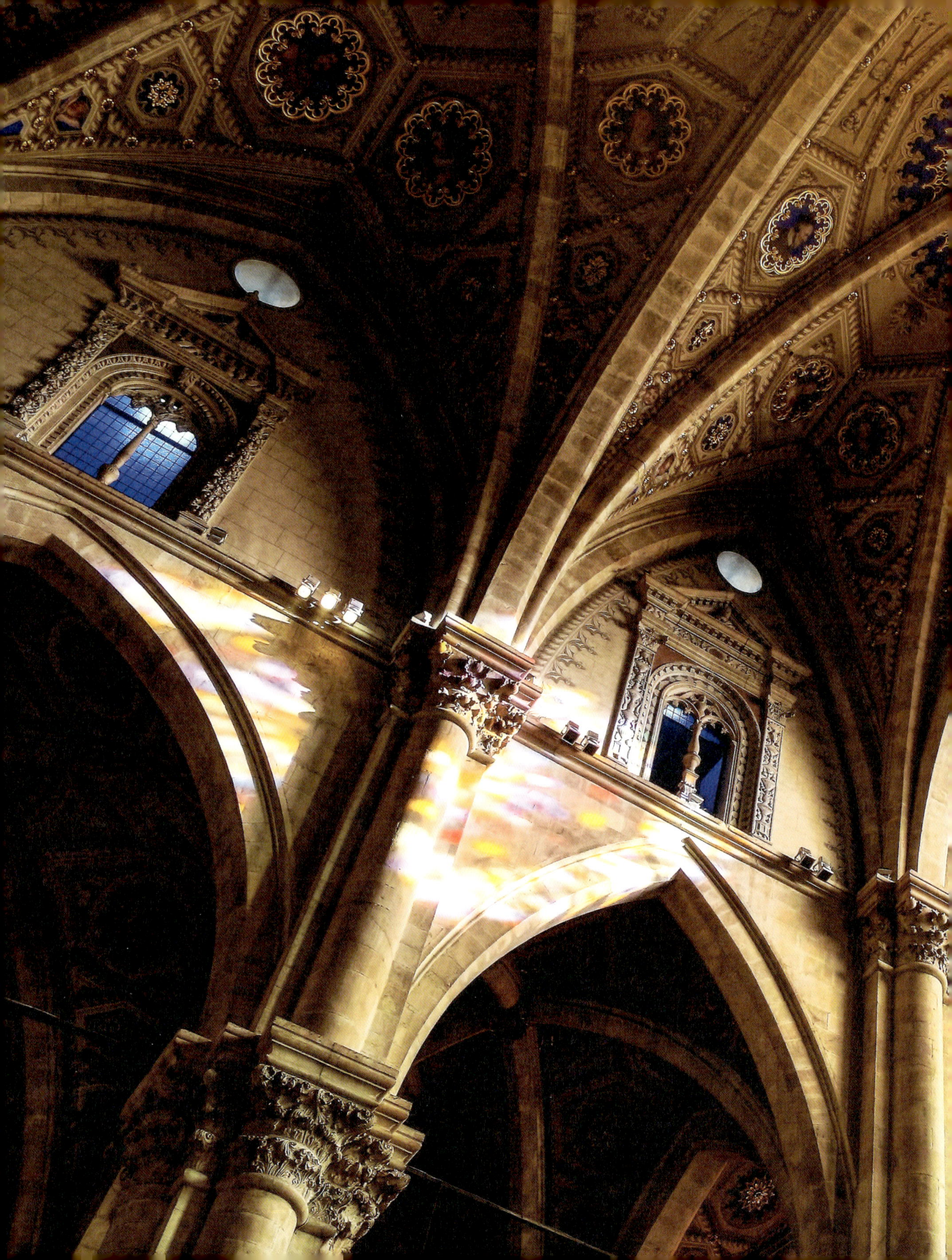

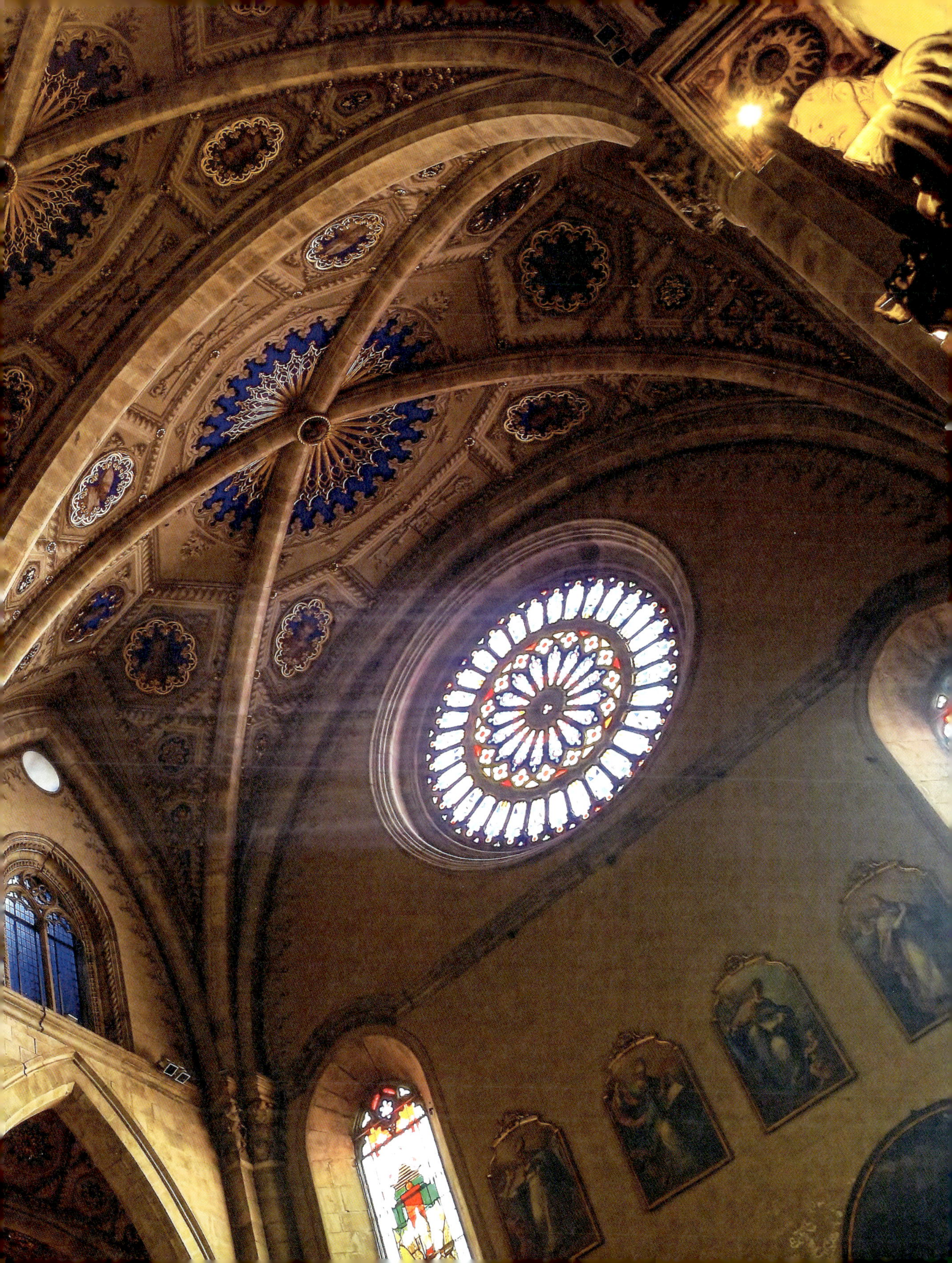

Willkommen im Mittelalter!

DER WANDEL
DER ROSE

VOM DEKADENTEN LUSTSINNBILD WIRD DIE ROSE
ZUM INBEGRIFF JUNGFRÄULICHER REINHEIT

Andere Zeiten, andere Werte: Als Symbol der Liebe erfuhr die Rose zwischen Antike und Mittelalter einen ausgeprägten Paradigmenwechsel. Wenn man im alten Rom an Rosen dachte, hatte man durch den betörenden Duft und die sinnliche Blütenpracht gleich das Bild ausschweifender Feste und dekadenter Opulenz im Sinn (siehe S. 103). Das frühe Christentum bediente sich dann einer bewährten und durchweg erfolgreichen Taktik: Anstatt heidnische Symbole und Brauchtümer zu verbieten, wurden diese einfach „gekapert" und uminterpretiert, sodass sie fortan zu den christlichen Wertvorstellungen passten. So wurden die römischen Saturnalien zu Weihnachten, dem Fest von Christi Geburt. Ähnlich erging es der Rose, die vom Sinnbild lasterhaften Treibens zum beliebten christlichen Symbol umgedeutet wurde und sich so weiterhin großer Beliebtheit erfreuen durfte. Dadurch war auch der Anbau von Rosen in vielen Klostergärten wieder „legitimiert" – ein Glück, dass die heilkräftige Blume nun quasi mit kirchlichem Segen ihre Wirkung tun durfte.

Die Blüte einer weißen Rose wurde im christlich geprägten Mittelalter zum Marienemblem erhoben, als Zeichen jungfräulicher Reinheit und keuscher Entsagung. In zahlreichen Kirchenliedern wird die Jungfrau Maria heute noch als „Rose ohne Dornen" besungen. Eine rote Rose wiederum, einst Symbol von Lust und Sinnlichkeit, wurde von nun an als Zeichen von Passion und hingebungsvoller Liebe zu Gott interpretiert. Die Darstellung von Rosenranken mit wilden Dornen gilt als Symbol von Martyrium und Hoffnung auf Erlösung – und so wird bisweilen auch Jesus' Dornenkrone mit roten Rosen geschmückt.

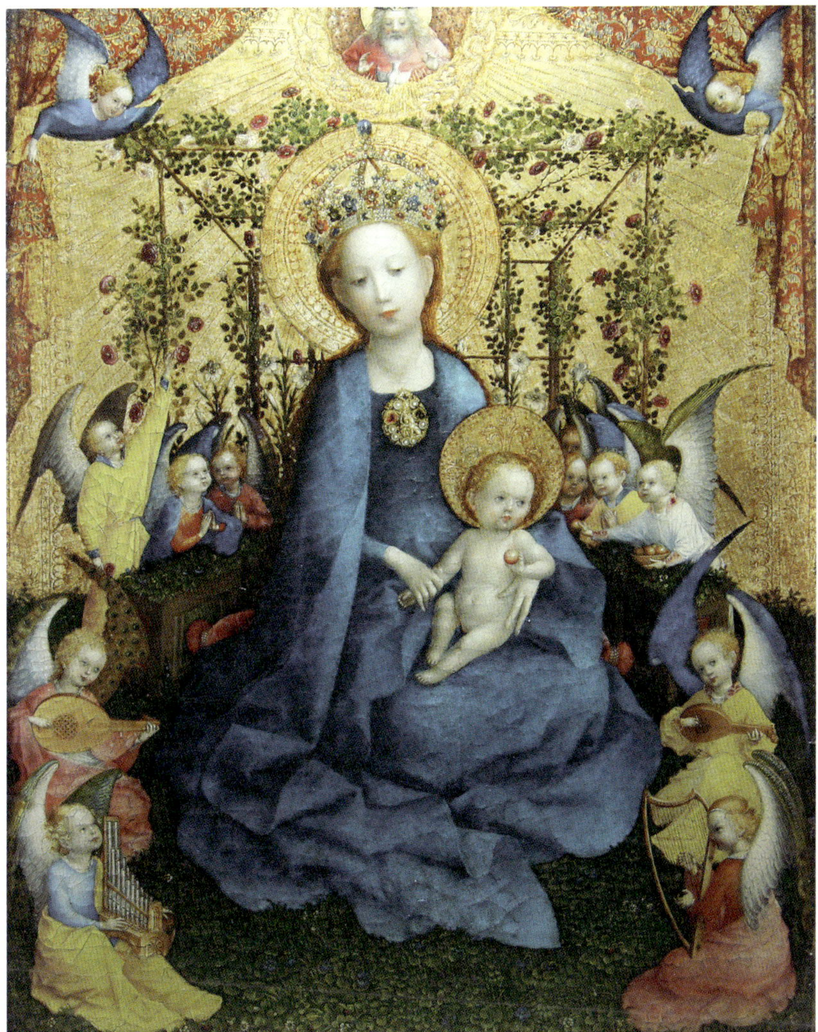

Madonna in the Rose Grove by Stefan Lochner († 1451). The roses behind her allude to the legend that roses were thornless before original sin.

Madonna im Rosenhag von Stefan Lochner († 1451). Die Rosen hinter ihr spielen auf die Legende an, dass Rosen vor der Erbsünde dornenlos waren.

Mittelalterliche Rosendarstellungen lassen die strenge Geometrie von fünf gleichmäßigen Blütenblättern erkennen: auch dies ein Indiz, dass das Pentagramm aus heidnischen Zauberkulten ganz bewusst in die christliche Nomenklatur aufgenommen wurde. So findet sich eines der ältesten Blumenmotive in Kunst und Architektur auch als stilisierte Rose in vielen mittelalterlichen Kirchen und Kathedralen wieder – von der Rosette als Schlussteil in Kreuzgewölben hin zu prachtvoll verglasten Fensterrosen. Über diesen Weg dürfte es die Rose auch zum Wappensymbol geschafft haben.

England's Wars of the Roses

ENEMIES UNDER THE ROSE EMBLEM

A DELICATE ROSE SYMBOLIZES ONE OF HISTORY'S MOST BITTER POWER STRUGGLES

There is a long tradition in England of bearing a rose on one's shield. Edward I selected a golden rose for his royal emblem, a well-known symbol of love and good fortune. For the queen of flowers to go down in history with the War of the Roses as a symbol of strife and contention seems rather unfair. How was it able to happen?

As with many disputes, the Wars of the Roses were preceded by long-running discord. Shortly before his death in 1376, Edward III had regulated his succession by decree, making it explicit that the throne would pass in the male line. Later, in 1399, when Henry IV forced his cousin Richard II to abdicate the throne, Henry based his own claim on the House of Lancaster, whose heraldic badge became known as the Red Rose. Claims to the throne in the female line of descent were simply overridden. The House of York—yes, the one with the White Rose emblem—was not about to permanently resign itself to these circumstances. But the tactic worked for some time; and meanwhile, Henry VI, king of England, had succeeded to the throne in tender infancy. Yet after the defeat in the Hundred Years' War, the fragile power structure began, around 1455, to permanently falter. The ambitious Richard of York had gained political influence as Lord Protector to King Henry VI (who in today's terms presumably was severely depressed), and Richard now saw the White Rose of York blossom into dominance. There followed a whirlwind of treachery and shifting loyalties, an unsettled back-and-forth on the throne. A quick rundown sounds like a sports commentator calling a rally at Wimbledon: Henry VI; Edward IV; back to Henry; right back to Edward, to York's White Rose; to that one's son; and then to

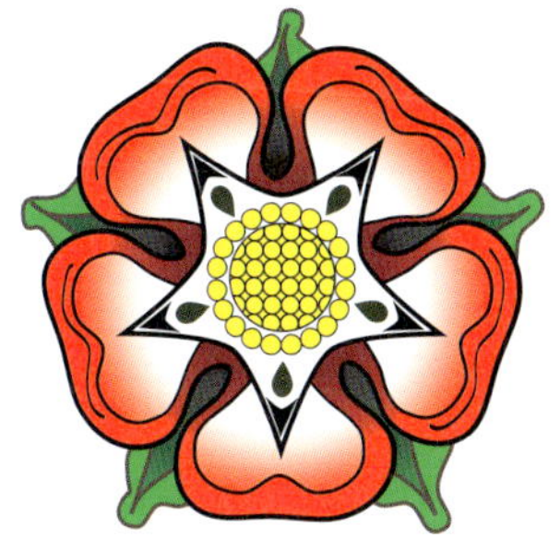

Richard III. *He* was ultimately defeated in 1485 at the Battle of Bosworth Field, by Henry VII, who concluded the War of the Roses with the foundation of the Tudor dynasty.

In such a drama, ripe for the stage, William Shakespeare found a great deal of material for his historical plays.

Whether the house of Lancaster itself carries the red rose on its shield is unclear to this day. It may be presumed to have been the rather clever idea of Henry VII, who through his marriage to Elizabeth of York had legitimized his claim to the throne via both lines of descent. He attributed the red rose to Lancaster to serve as an antipode to the white York rose, and from it he created "his" Tudor rose. Henceforth, the emblem of the red rose enclosing a white floral wreath was to unite the opposing lines.

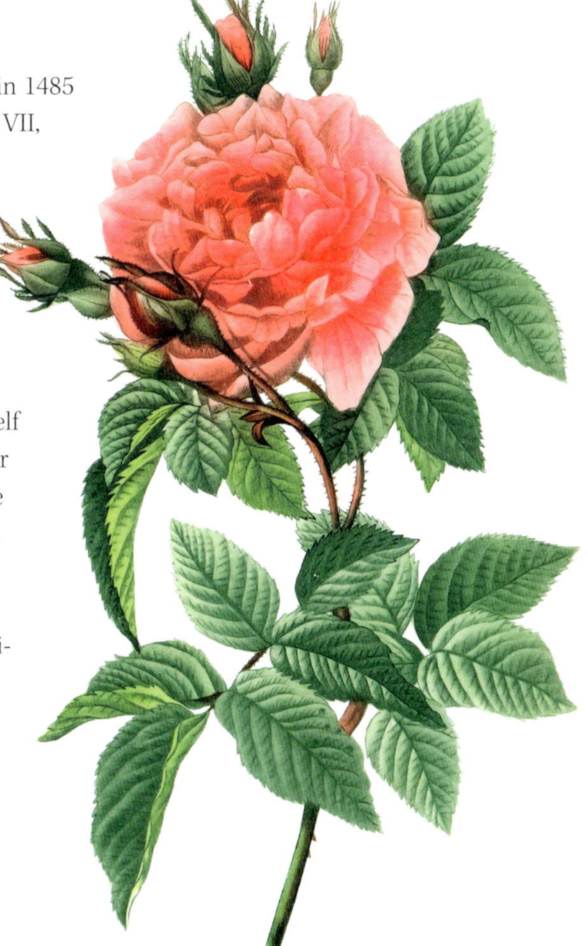

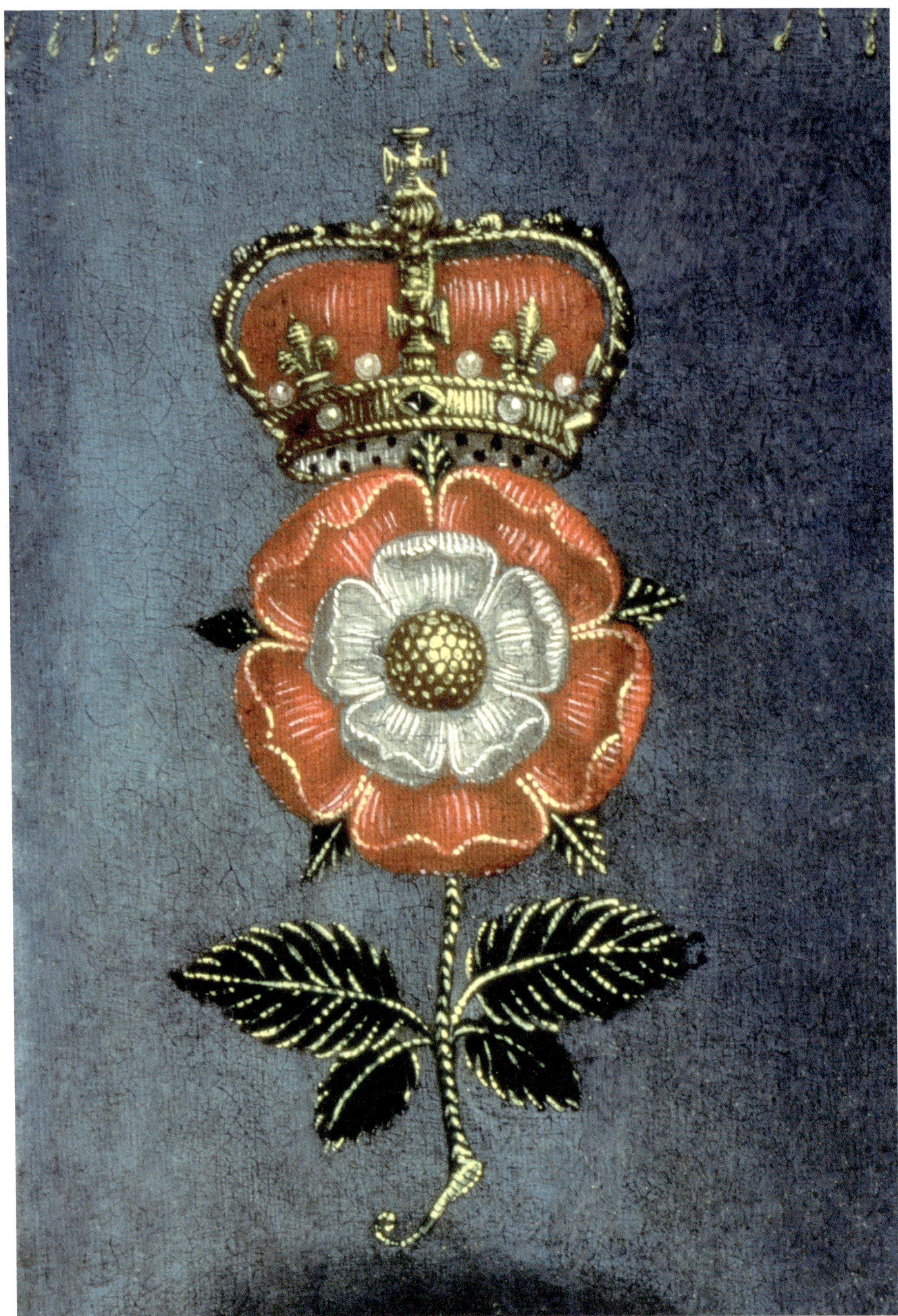

Rosenkriege in England

VERFEINDET UNTERM ROSENEMBLEM

EINE ZARTE BLÜTE ALS SINNBILD FÜR EINEN DER ERBITTERTSTEN MACHTKÄMPFE DER GESCHICHTE

Eine Rose im Schilde zu führen – das hat in England eine lange Tradition: Bereits Edward I. wählte als royales Emblem die goldene Rose, ein bekanntes Symbol für Liebe und Glück. Dass mit den Rosenkriegen die Königin der Blumen dann wiederum als Sinnbild für Zank und Streit in die Geschichte eingehen sollte, erscheint schon etwas unfair. Aber wie kam es nun dazu?

Mit den Rosenkriegen ist es wie bei vielen Streitigkeiten: Der Zwist schwelte lange, bevor offener Konflikt ausbrach. Edward III. hatte seine Nachfolge kurz vor seinem Tod im Jahr 1376 per Erlass geregelt, entsprechend wurde der Thron nun explizit in männlicher Linie vererbt. Als Henry IV. 1399 seinen kinderlosen Cousin Richard II. zur Abdankung zwang, überging er den Erbanspruch der Nachfahren von Edwards zweitältestem Sohn Lionel, der nur in weiblicher Linie weitergegeben wurde. Mit Henry IV. bestieg der Nachkomme von Edwards drittem Sohn John, Duke of Lancaster, den Thron – das Haus Lancaster, welches später als „Rote Rose" bekannt wurde, war geboren.

Eine Zeit lang ging die Taktik noch auf, mittlerweile war Henry VI. im zarten Babyalter zum Thronerben geworden. Doch nach der Niederlage im Hundertjährigen Krieg kam das fragile Machtgefüge um 1455 endgültig ins Wanken: Der ehrgeizige Richard von York hatte als „Protektor"

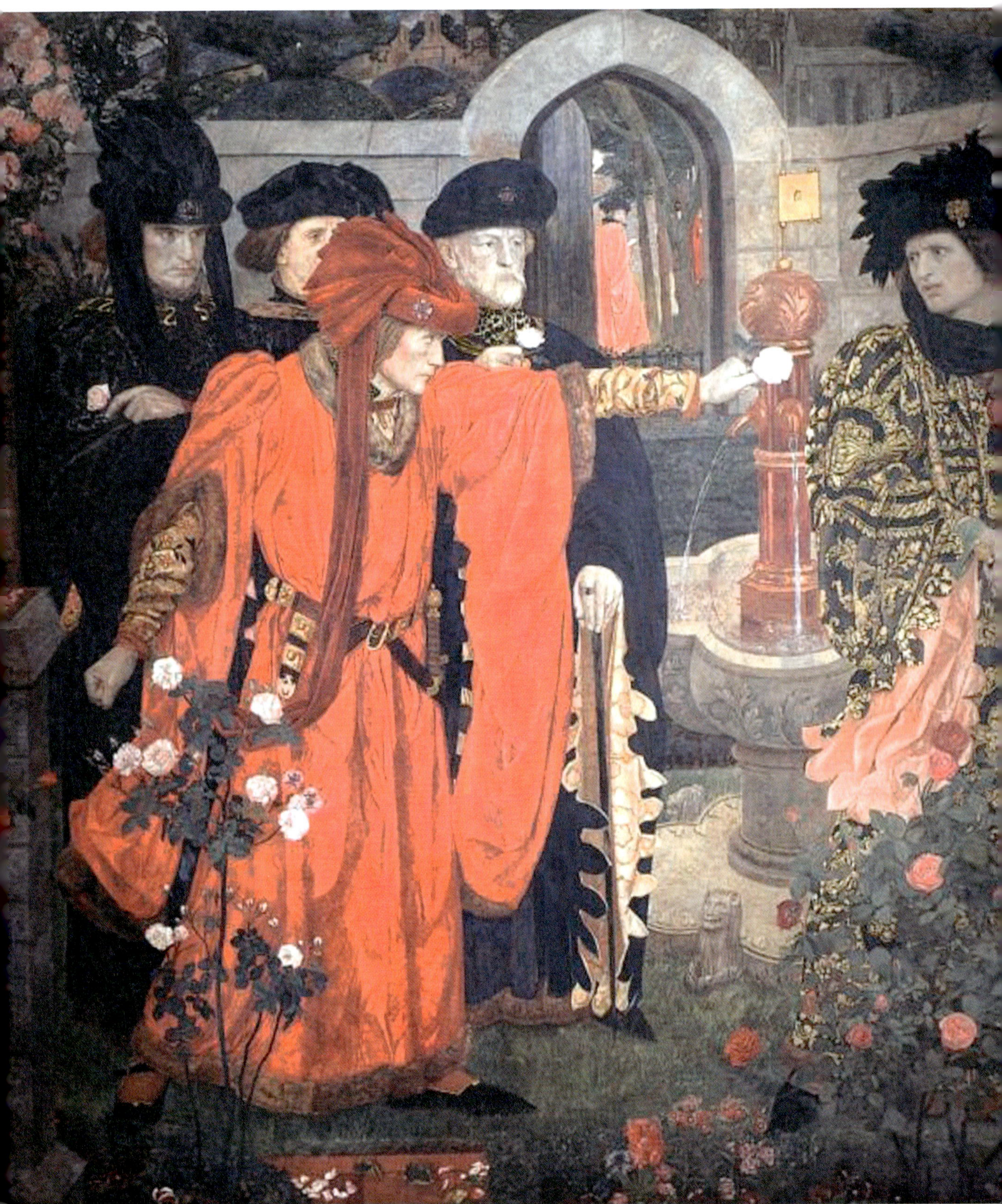

des aus heutiger Sicht vermutlich schwer depressiven Königs Henry VI. an politischem Einfluss gewonnen und sah nunmehr die weiße Rose aus York zur Herrschaft erblühen. Es folgte ein Wirbel aus Verrat und wechselnden Loyalitäten, mit wildem Hin und Her auf dem Thron. Im Schnelldurchlauf liest sich das wie ein Ballwechsel in Wimbledon: von Henry VI. zu Edward IV., dann wieder zurück zu Henry – und gleich wieder hin zur York'schen weißen Rose mit Edward, dann dessen Sohn, dann Richard III. Diesen besiegte Henry VII. schließlich 1485 in der Schlacht von Bosworth und beendete mit der Gründung der Tudor-Dynastie die Rosenkriege. Ein so bühnenreifes Geschichtsdrama, dass William Shakespeare darin jede Menge Stoff für seine Historienstücke fand!

Ob nun das Haus Lancaster selbst schon die rote Rose im Schilde führte, ist heute noch unklar: vermutlich eher ein cleverer Einfall Henrys VII., der durch seine Heirat mit Elizabeth von York seinen Herrschaftsanspruch über beide Erblinien legitimierte. Als Gegenpart zur weißen York-Rose schrieb er Lancaster die rote Rose zu – und schuf daraus „seine" Tudor-Rose: Fortan sollte das Emblem der roten Rose, die einen weißen Blütenkranz umschließt, die verfeindeten Linien vereinen.

LEFT PAGE | A stylized representation of the feuding houses from a mural at the Palace of Westminster

LINKE SEITE | Wandgemälde im Palace of Westminster – eine stilisierte Darstellung der verfeindeten Häuser

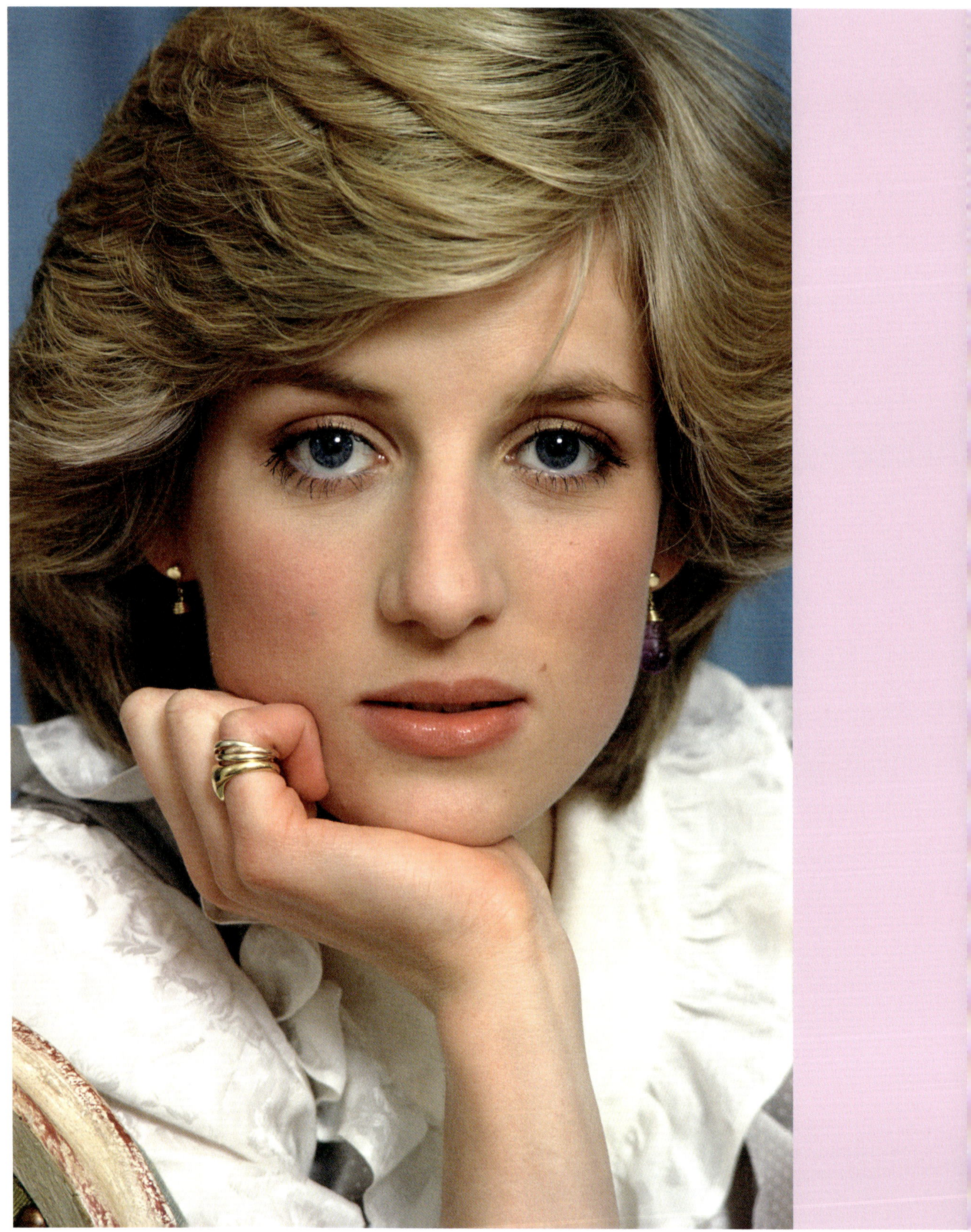

"The English Rose"

BEAUTY
IN BLOOM

THE NATIONAL FLOWER OF ENGLAND STANDS FOR GRACE AND BEAUTY – AND HAS PROMINENT REPRESENTATIVES

Not only has the rose been England's national flower since the thirteenth century, the English rose also has a long tradition as the symbol of a natural and authentic beauty. By the time Elton John transformed his wistful ballad *Candle in the Wind* into a final *Goodbye, England's Rose* for Lady Di's funeral, the former princess of Wales had already enshrined her rose moniker as a synonym for beauty and grace far beyond the British Isles.

To look at a picture of Diana is to immediately recognize an ideal of natural beauty in bloom: flawless complexion, pale and elegant; light soft-pink flush of the cheeks, as though just back from a stroll through a spacious park or along a lakeside promenade; lips full and slightly reddened, as if to tell of a secret kiss; her eyes bright and clear, her aspect lowered, reserved, with a shy upward gaze—same as the natural beauty of the English rose has been depicted for centuries in painting and literature. Blond, red, or brunette, the beauty ideal is not tied to a particular hair color. It is no surprise that Diana's daughter-in-law, Catherine, now princess of Wales, is following worthily in the footsteps of presumably

LEFT PAGE | Diana, Princess of Wales at home in Kensington Palace (1983)

LINKE SEITE | Diana, Princess of Wales, zu Hause im Kensington Palace (1983)

the most famous English rose. Actors such as Keira Knightley, Kate Winslet, Emma Watson, and Rosamund Pike also tend to receive blossoming-rosy-beauty nicknames. Incidentally, Japan has a very similar floral ideal of female loveliness; there, however, the *yamato nadeshiko* is an allusion to a delicate carnation blossom.

The concept of the "English Rose" certainly seems to have royal origins: ever since England's queen Elizabeth I chose the rose as her emblem, people have associated the delicate rose, as a symbol of virgin beauty, with the once-cherished ideal of the beauty of an almost ethereally fair complexion. But the ideal encompasses far more than the merely superficial. As its portrayal in film and literature implies, it is also the essence of a true English lady: genteel reserve; gentle and patient; gracious; respectful; virtuous; or in short, a flock of adjectives to match to the passivity and tolerance of the female role model that for centuries had set the tone. Will the ideal of the English rose wilt under these changing times?

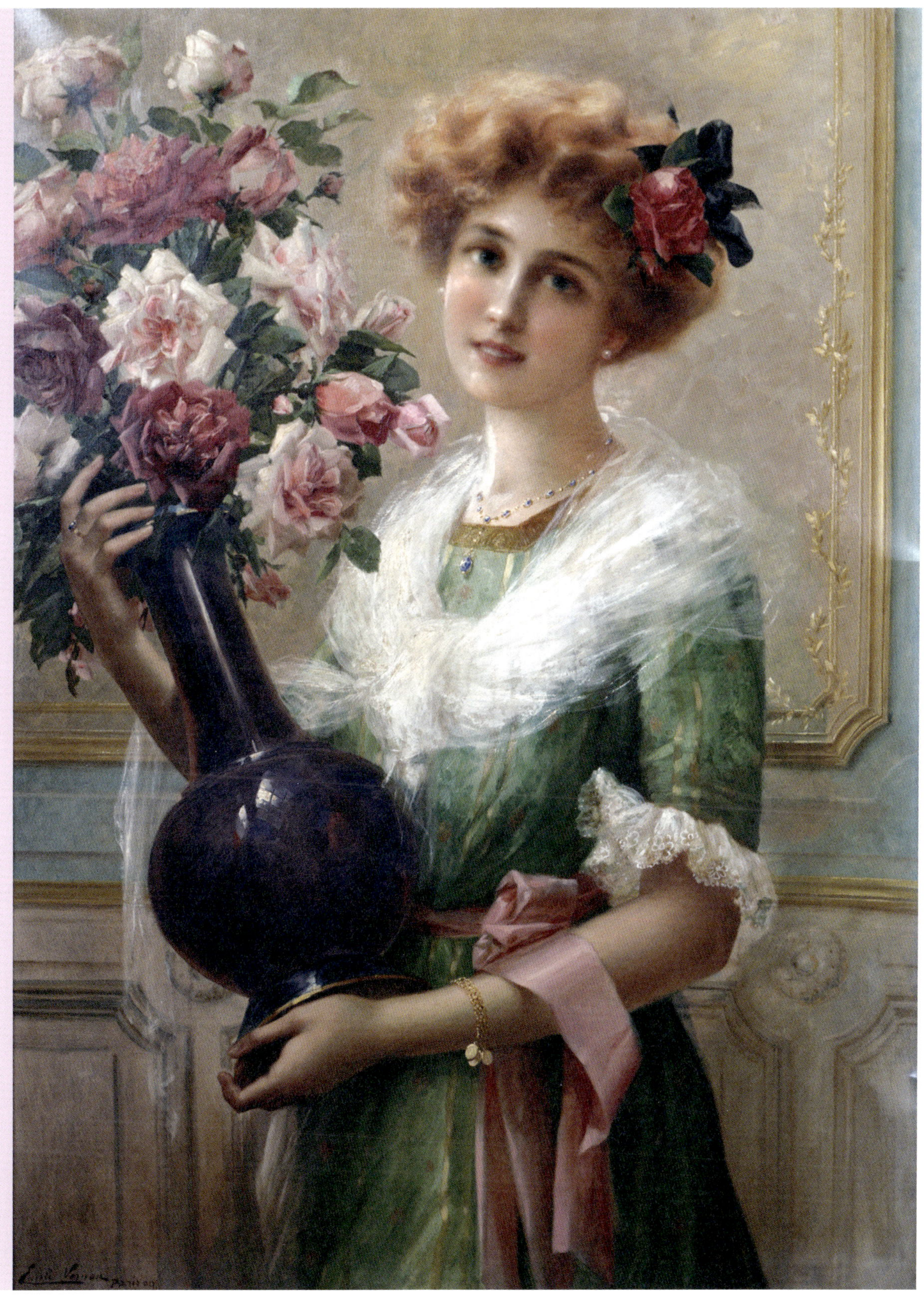

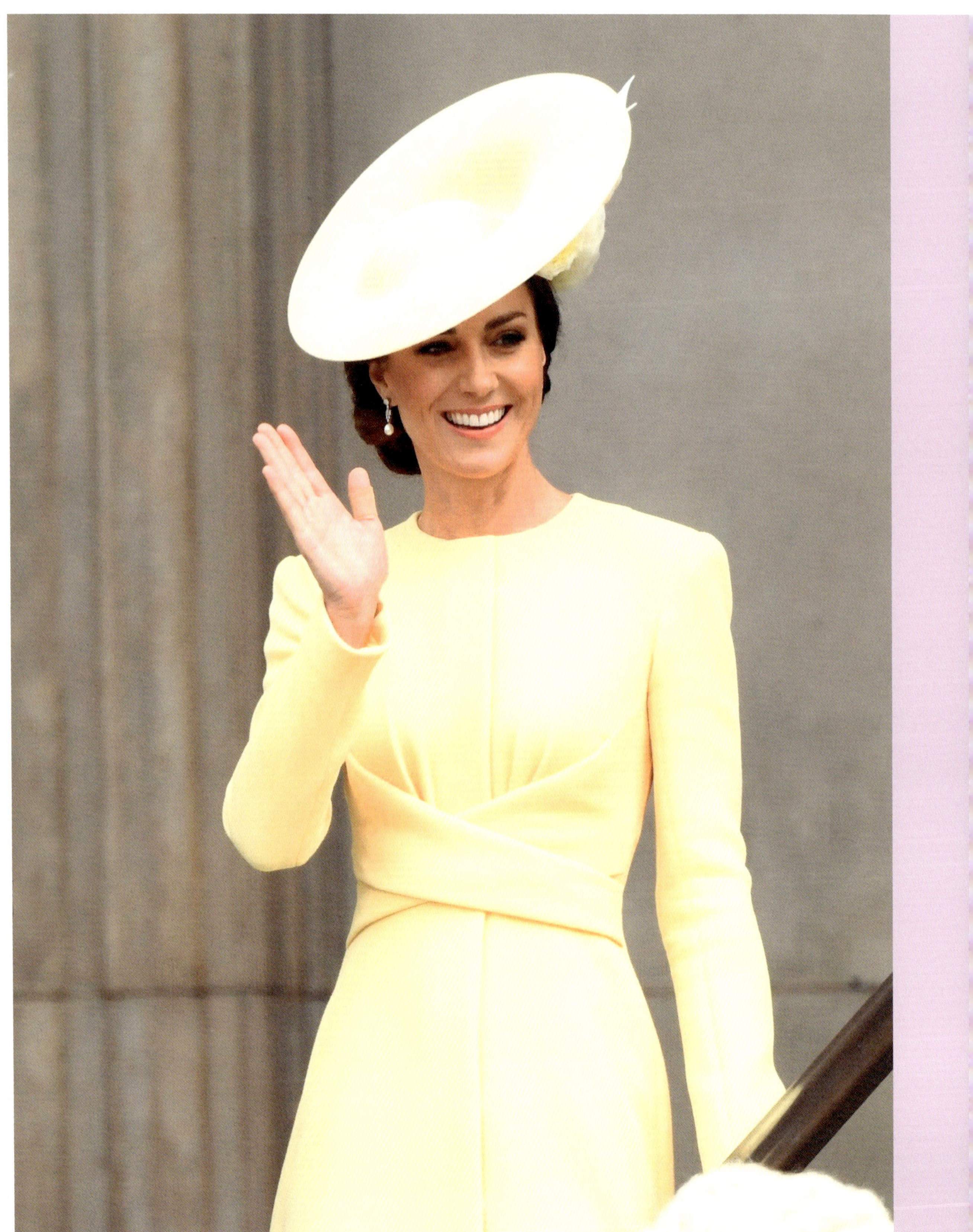

Die „Englische Rose"

BLÜHENDE SCHÖNHEIT

DIE NATIONALBLUME ENGLANDS STEHT FÜR ANMUT UND GRAZIE – UND HAT PROMINENTE VERTRETERINNEN

Die Rose ist nicht nur seit dem 13. Jahrhundert Englands National-blume – auch als Symbol für eine natürliche, unverfälschte Art der Schönheit hat die *English Rose* eine lange Tradition. Spätestens als Elton John nach dem Tod Lady Dis seine wehmütige Ballade „Candle in the Wind" in ein letztes „Goodbye, England's Rose" verwandelte, wurde mit der einstigen Prinzessin von Wales auch außerhalb der Britischen Inseln die Rose zum Synonym von Schönheit und Anmut.

Und wer Dianas Bild vor Augen hat, weiß auch gleich, wie das Ideal der blühenden natürlichen Schönheit aussehen soll: ein makelloser Teint von vornehmer Blässe, die Wangen überhaucht von sanftem Rosa, als ob sie gerade von einem Spaziergang durch weitläufige Parks oder vom Flanieren entlang der Seepromenade zurückgekehrt sei. Die Lip-pen voll und leicht gerötet, gerade so, als erzählten sie von einem heim-lichen Kuss. Die Augen klar und hell, aus dem zurückhaltend gesenkten Blick ein wenig schüchtern emporschauend – so, wie seit Jahrhunderten die natürliche Schönheit der *English Rose* in Malerei und Literatur ge-zeichnet wird.

LEFT PAGE | Catherine, Princess of Wales, London (2022)

LINKE SEITE | Catherine, Princess of Wales, London (2022)

Ob blond, rötlich oder brünett: Auf eine bestimmte Haarfarbe ist das Schönheitsideal nicht festgelegt. So erstaunt es nicht, dass Dianas Schwiegertochter Catherine mit ihrem natürlichen Look würdig in die Fußstapfen der wohl berühmtesten *English Rose* tritt. Auch Schauspielerinnen wie Keira Knightley, Kate Winslet, Emma Watson oder Rosamund Pike werden gerne mit der Rosenallegorie beschrieben. Ein ganz ähnliches florales Idealbild des weiblichen Liebreizes gibt es übrigens auch in Japan: Dort spielt die *yamato nadeshiko* allerdings auf eine zarte Nelkenblüte an.

Das Konzept der *English Rose* scheint durchaus royale Ursprünge zu besitzen: Seit Englands Königin Elizabeth I. (1533–1603) sich die Rose als Emblem erwählte, verbindet man die zarte Blüte mit dem damals geschätzten Schönheitsideal eines fast ätherisch hellen Teints. Das Idealbild der *English Rose* umfasst aber weit mehr als nur Äußerlichkeiten; vielmehr ist es, wie auch die Darstellung in Film und Literatur impliziert, das typische Bild einer echten englischen Lady: vornehm zurückhaltend, sanft und geduldig, graziös und anmutig, respektvoll, tugendhaft – kurzum, eine Menge Adjektive, die dem jahrhundertelang geforderten passiv-duldsamen weiblichen Rollenbild entsprechen. Ob mit dem Wandel der Zeiten bald auch das Ideal der *English Rose* Geschichte sein wird?

RIGHT PAGE | *An English Rose* (1892), portrait by the British painter Eva Hollyer

NEXT PAGE | More than a century later, rose-breeder David Austin introduced the nostalgic-romantic 'A Shropshire Lad'.

RECHTE SEITE | Porträt *An English Rose* (1892) der britischen Malerin Eva Hollyer

NÄCHSTE SEITE | Der Rosenzüchter David Austin hat die nostalgisch-romantische „A Shropshire Lad" gut 100 Jahre später vorgestellt.

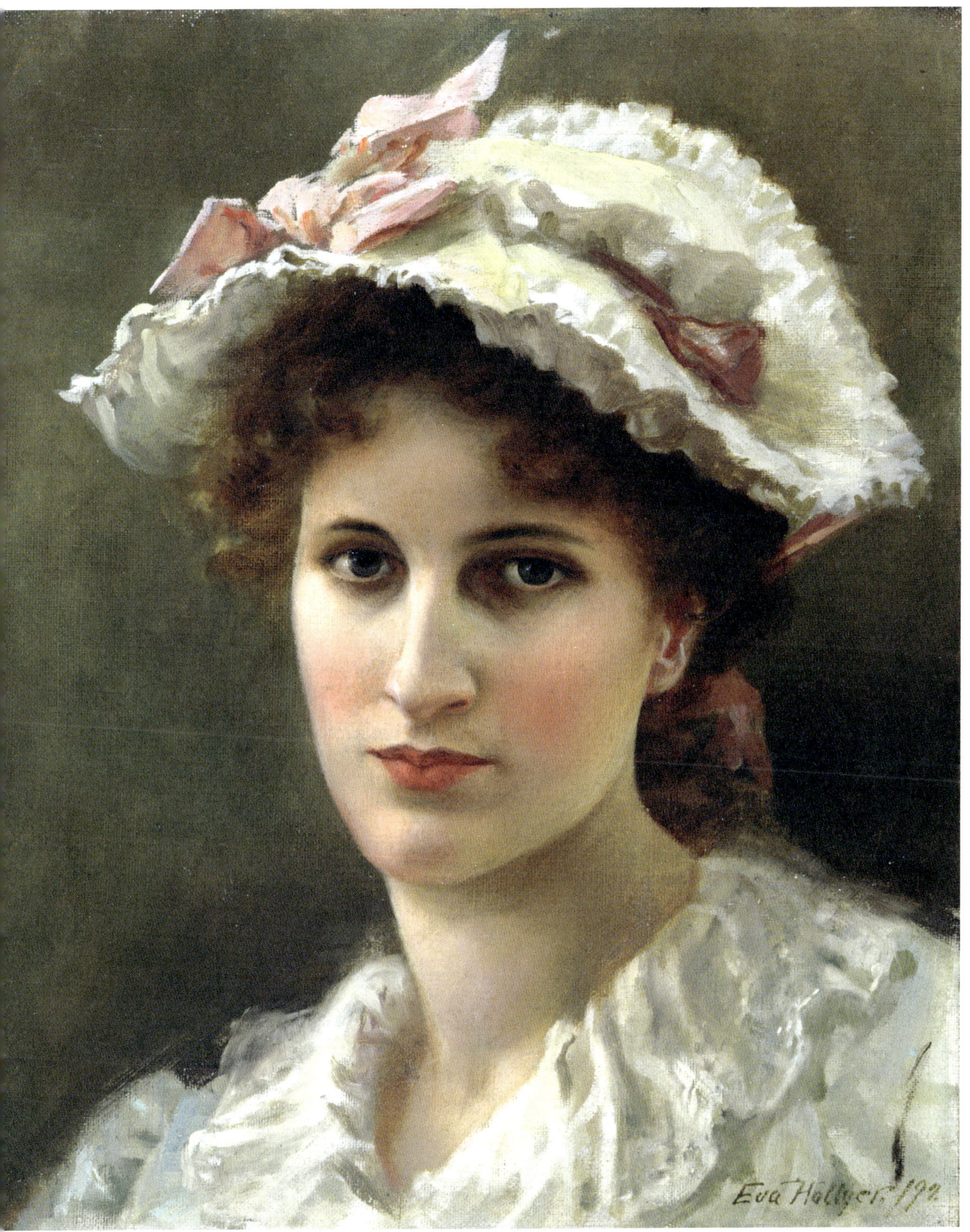

Eva Hollyer /92

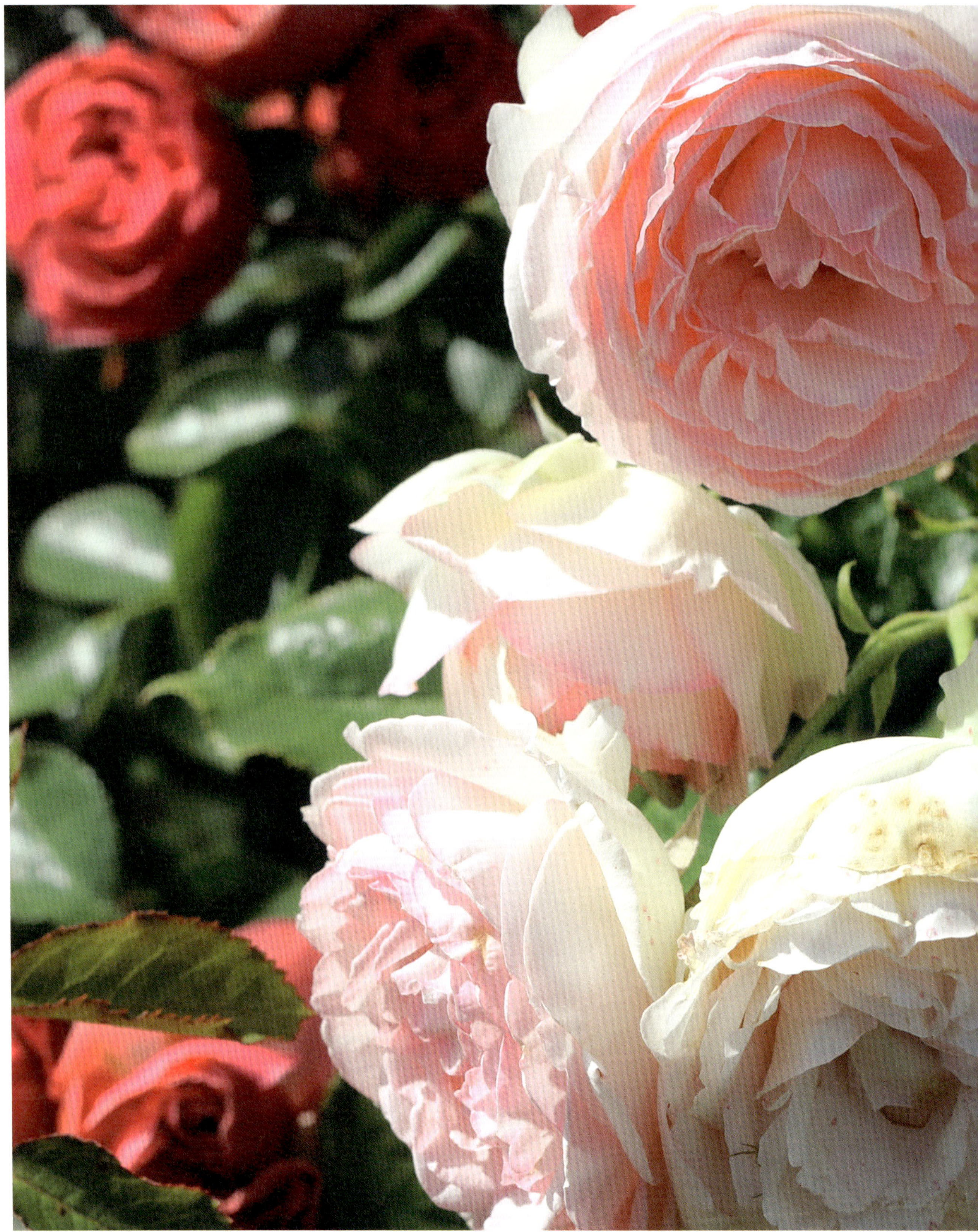

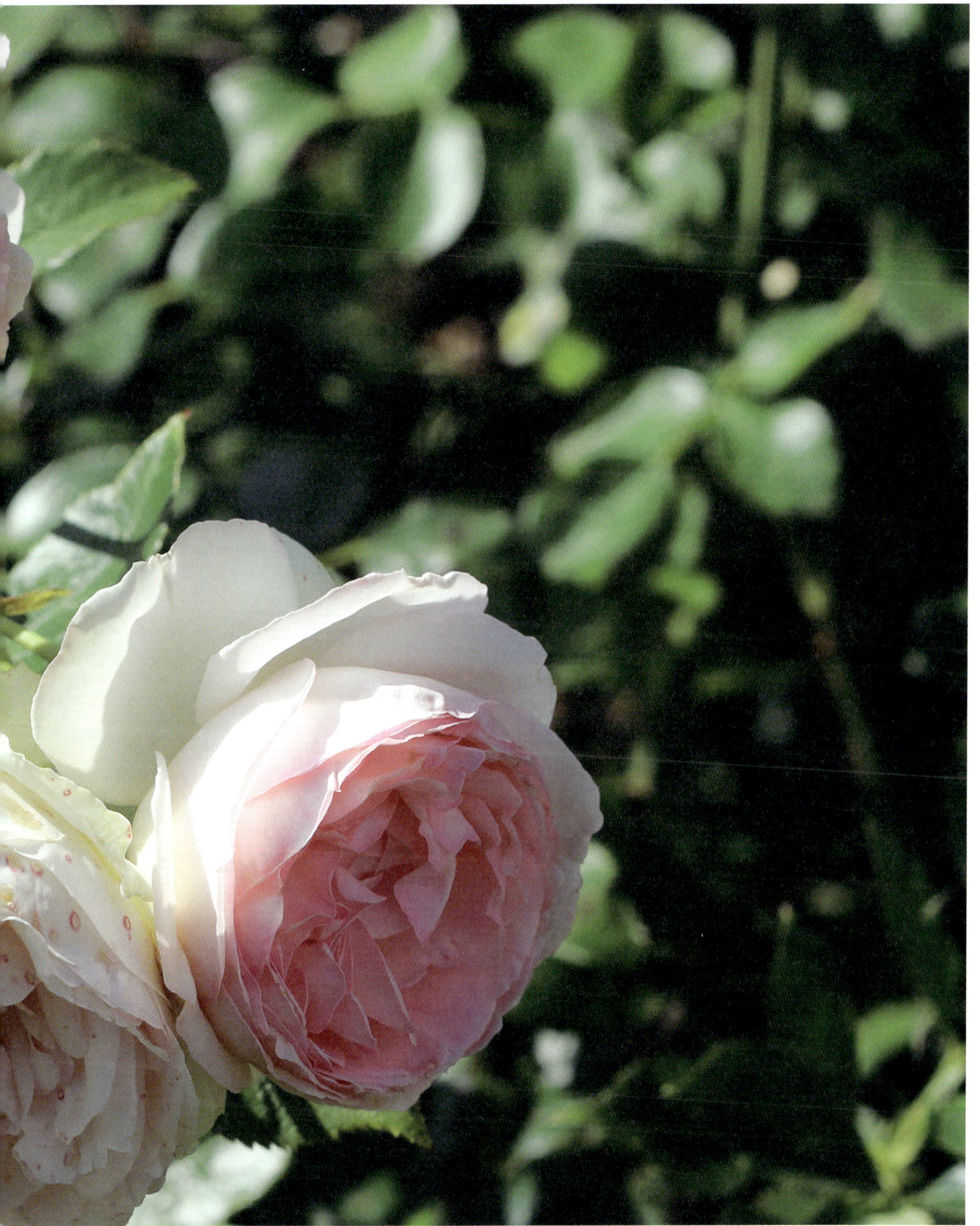

GROWING ROSES AT THE BEHEST OF THE EMPEROR

CHARLEMAGNE, ROSE-FRIEND?

For Europe's most powerful ruler of old, this admittedly will not be the first by-name that springs to mind. But behind Charles's enthusiasm for roses lay not a poetic-romantic devotion but rather foresight in providing for his court and subjects.

Enormous power, enormous duties: To unify his empire and consolidate his rule, Charlemagne traveled throughout his life, going to and from the more than one hundred imperial palaces of his vast realm, always with an extensive entourage of up to a thousand people, who of course had to be

Statue of Charlemagne in Frankfurt

Statue Karls des Großen in Frankfurt am Main

provided for. Charles, clearly a pragmatist, recognized that the glaring lack of food stores he would sometimes encounter locally had the potential to undermine his rule. In 812, he issued a decree, the *Capitulare de villis*, to remedy the situation, regulating exactly what was to be available on his estates, from agriculture to livestock, and in what quantity, to provide for the court. The decree was also supposed to ensure that his subjects would have access to basic supplies even when the emperor was away. While it may seem surprising from our point of view for the king to be the provider, it was smart domestic policy. Full bellies are less likely to plot against the ruler.

The *Capitulare de villis* describes in detail what was to be grown on the farms in each palatinate: fruits, vegetables, medicinals, and other plants, on the model of monastic herb gardens. This is where the rose comes into play—specifically, the dog rose, *Rosa canina*. Charles, inquisitive, valued the medicinal benefits of rose oil, rosebuds, rose petals, and rose hips. Monastic medicine even then knew of many uses for the powerful benefits of the rose, be it for ailments of the eyes or lungs, abscesses, common colds, or digestive problems. Today it is scientifically proven that the dog rose that Charles so highly esteemed contains (for example) high levels of vitamin C and tannins, confirming not only the ruler's shrewd instincts but also the breadth of experience of the medieval herbalists.

NEXT PAGE | Rose hips, whose health-giving properties were known even in the emperor's day

NÄCHSTE SEITE | Hagebutten, um deren Wert für die Gesundheit man schon zu Kaisers Zeiten wusste

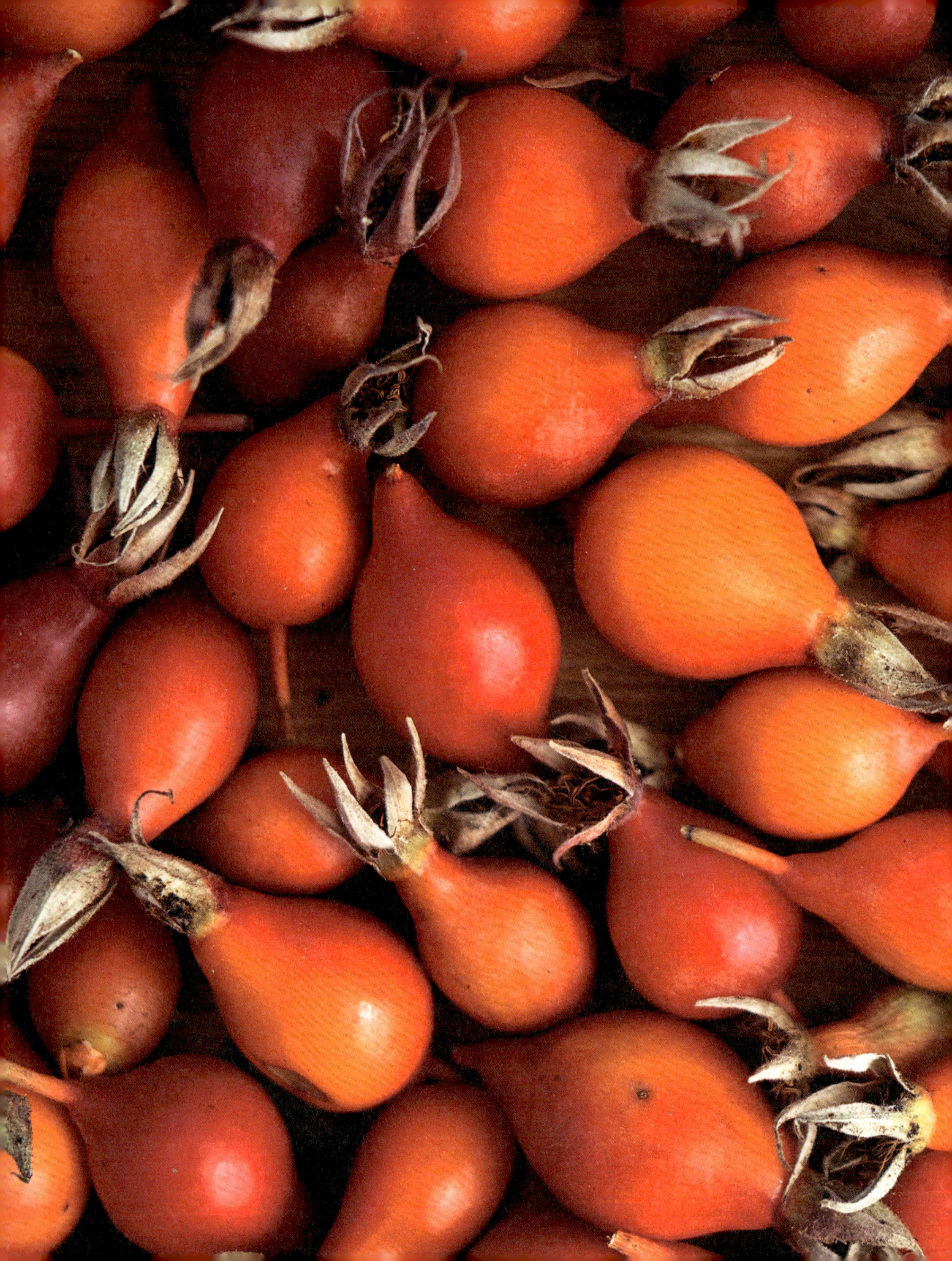

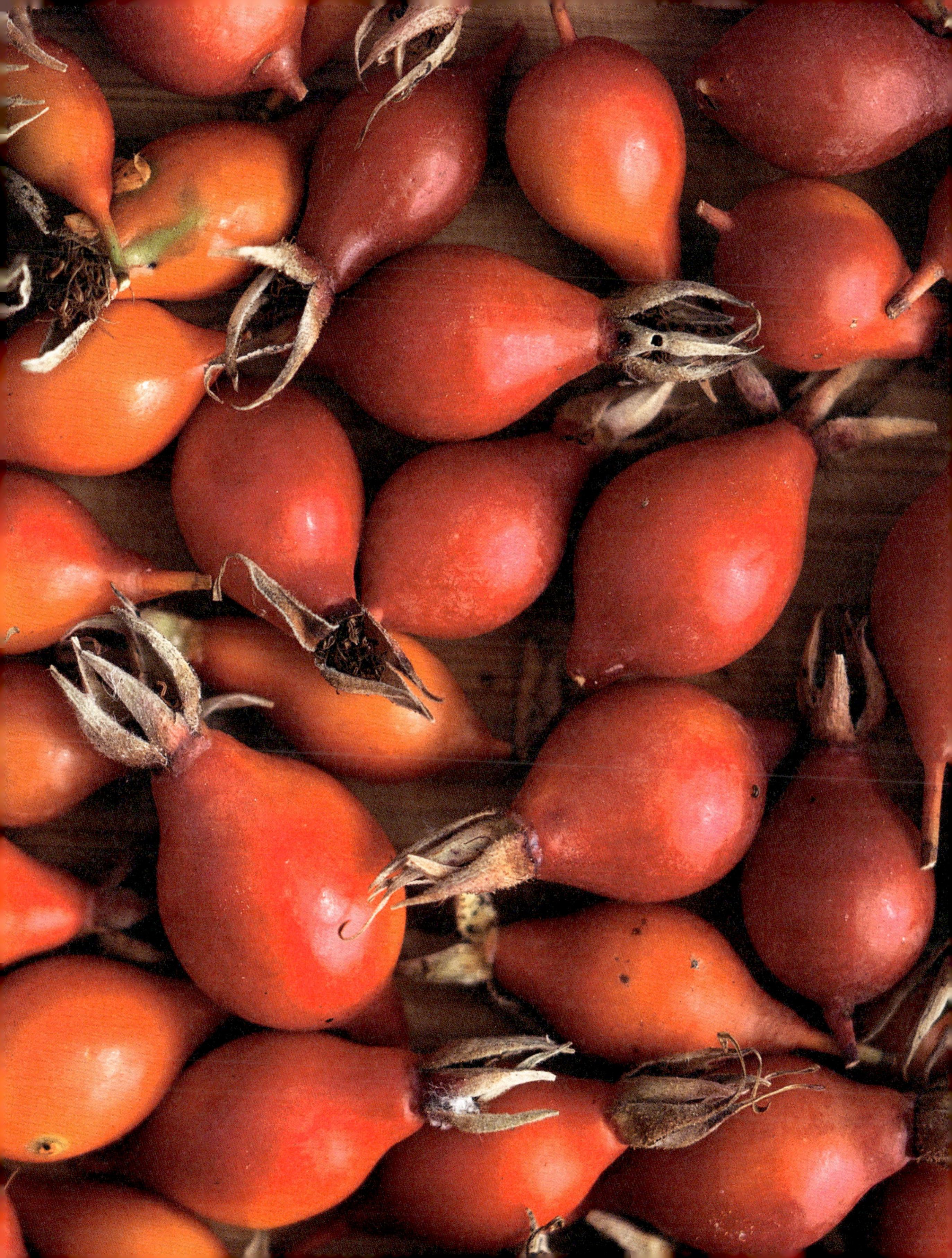

Le Petit Journal

ADMINISTRATION
61, RUE LAFAYETTE, 61

Les manuscrits ne sont pas rendus

On s'abonne sans frais
dans tous les bureaux de poste

5 CENT.

SUPPLÉMENT ILLUSTRÉ

5 CENT.

23me Année — ++ — Numéro 1.134

DIMANCHE 11 AOUT 1912

ABONNEMENTS

	SIX MOIS	UN AN
SEINE et SEINE-ET-OISE..	2 fr.	3 fr. 50
DÉPARTEMENTS............	2 fr.	4 fr. »
ÉTRANGER	2 50	5 fr. »

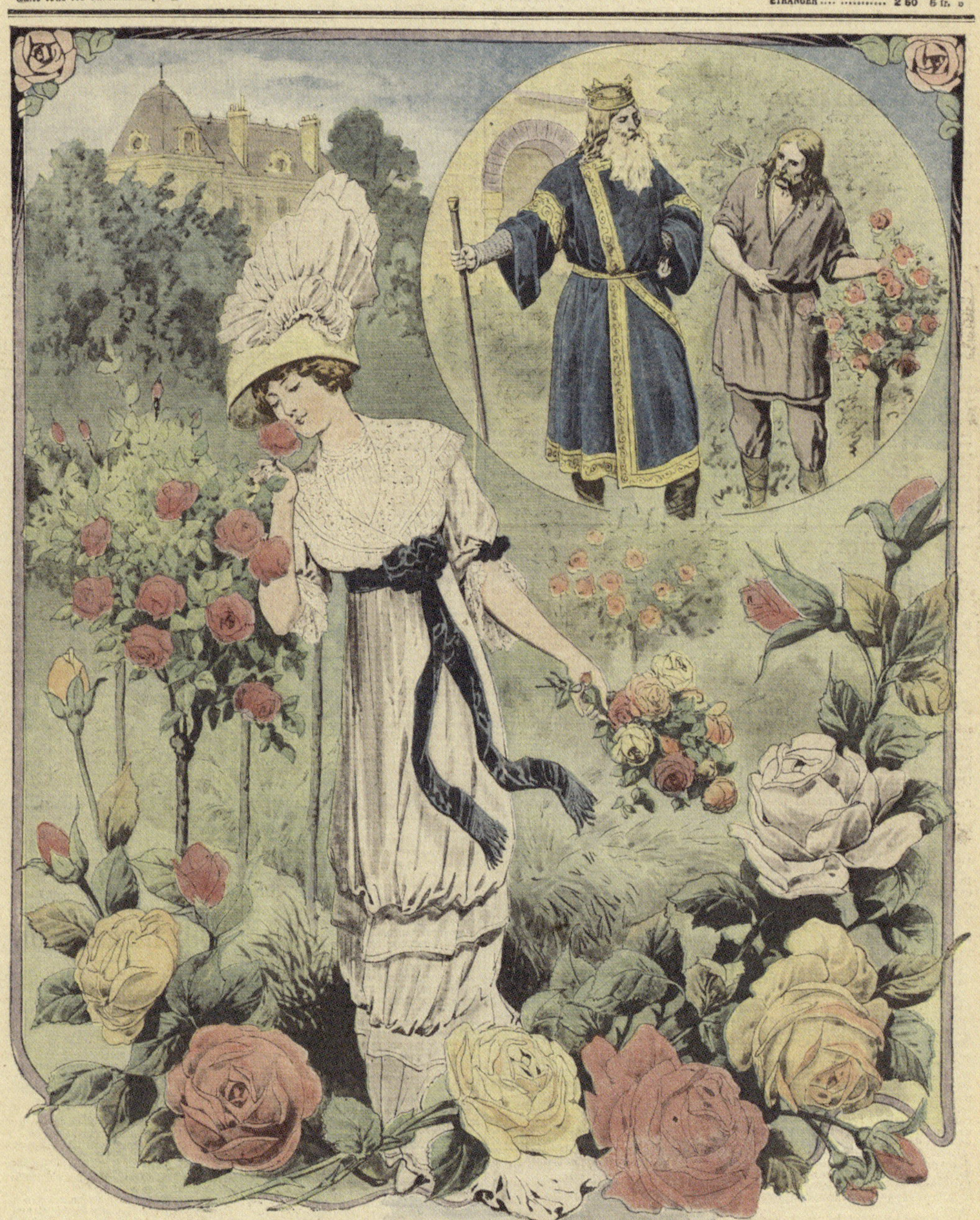

LE JUBILÉ DE LA ROSE

Il y a juste onze siècles que l'empereur Charlemagne prit un édit
pour propager la culture du rosier au pays des Francs

9. Jahrhundert

ROSENANBAU AUF KAISERS GEHEISS

KARL DER GROSSE: EIN ROSENFREUND?

Zugegeben, vielleicht nicht der erste Beiname, der uns zum einst mächtigsten Herrscher Europas einfällt. Doch hinter Karls Begeisterung für Rosen steckt nicht poetisch-romantische Hingabe, sondern vorausschauende Verantwortung für Hofstaat und Untertanen.

Große Macht, große Verpflichtungen: Im Bestreben, sein Imperium zu einen und seine Herrschaft zu festigen, reist Karl der Große zeitlebens kreuz und quer zu den über 100 Kaiserpfalzen seines Riesenreichs, immer mit einem umfangreichen Hofstaat von bis zu 1000 Menschen im Schlepptau, der natürlich versorgt werden muss. Der offenbar sehr pragmatisch veranlagte Karl erkennt, dass die teils eklatanten Missstände in der Versorgung vor Ort seine Herrschaft gefährden könnten. Abhilfe schaffen soll sein Erlass *Capitulare de villis* aus dem Jahr 812: In der Landgüterverordnung wird von Ackerbau bis Viehzucht exakt geregelt, was auf den Hofgütern jeweils in welcher Menge für die Bewirtung des Hofstaats vorhanden sein muss. Und auch über den kaiserlichen Besuch hinaus soll die Grundversorgung seiner Untertanen dadurch gesichert sein. Der aus unserer Sicht vielleicht erstaunlich fürsorgliche Ansatz war natürlich auch innenpolitisch ein kluger Schachzug – gut gefüllte Bäuche planen seltener einen Aufstand gegen den Herrscher …

LEFT | In 1912, the Parisian newspaper *Le Petit Journal* proclaimed a rose jubilee in reference to Charlemagne's horticultural guidance from the year 812.

LINKE SEITE | Im Jahr 1912 rief die Pariser Tageszeitung *Le Petit Journal* das „Jubiläum der Rose" aus, bezogen auf die gärtnerischen Vorgaben Karls des Großen aus dem Jahr 812.

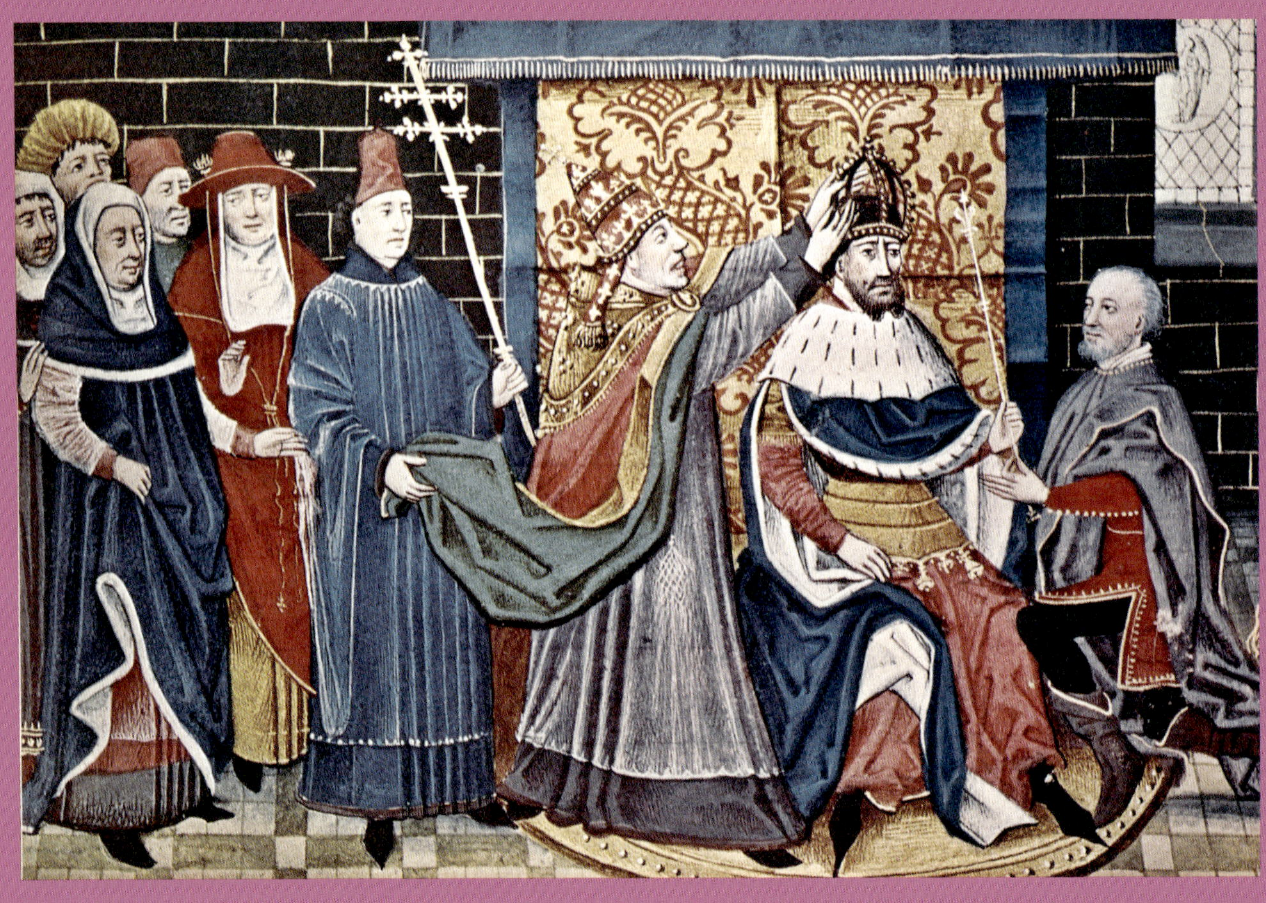

Charlemagne was crowned Roman
Emperor by Pope Leo III on December 25
in the year 800.

Karl der Große wurde am 25. Dezember des
Jahres 800 n. Chr. von Papst Leo III. zum
römischen Kaiser gekrönt.

Im *Capitulare de villis* ist detailliert beschrieben, was auf den Höfen jeder Pfalz angebaut werden soll: Obst, Gemüse sowie Heilkräuter und Heilpflanzen, inspiriert vom Vorbild klösterlicher Kräutergärten. Hier kommt nun die Rose ins Spiel, genauer gesagt *Rosa canina,* die Hundsrose. Karl, der Wissenschaft gegenüber stets aufgeschlossen, schätzt den medizinischen Nutzen von Rosenöl, Rosenknospen, Rosenblättern und Hagebutten. Die klösterliche Heilkunde kannte damals bereits viele Einsatzgebiete für die heilkräftige Rose, ob bei Augenleiden, Abszessen, Erkältungen, Lungenleiden oder Verdauungsbeschwerden. Heute ist für Karls geschätzte Hundsrose der hohe Gehalt zum Beispiel an Vitamin C und Gerbstoffen wissenschaftlich nachgewiesen – was den guten Instinkt des Herrschers und die reiche Erfahrung der mittelalterlichen Kräuterheilkundigen bestätigt.

Nineteenth century

THE ROSE PARADISE MALMAISON

EMPRESS JOSÉPHINE AND THE ROSES

Napoleon Bonaparte is said to have passionately loved his first wife, Joséphine de Beauharnais; but apparently—apart from Napoleon—her passion was roses. Although the French imperial couple resided at the Tuileries, Joséphine purchased (with generous financial support from Napoleon) the small chateau Malmaison, south of Paris, and remodeled it to suit her tastes. The wonderful gardens were the crowning glory; in them, the flower-lover collected rare plants from all over the world.

No reason is documented for the empress's particular fondness for roses. (Maybe it was because her third given name was Rose.) In any case, Joséphine's goal was to collect every known rose variety at Malmaison, at any cost. Some of the most renowned rose growers (even then) were in England—with which France was, stupidly and very persistently, at war. But love is love, and so to quell Joséphine's passion for roses, Napoleon hesitated not stop at boarding and searching enemy ships. The outcome was that many plants from England's colonies wound up at Malmaison. The *pièce de résistance* was the coveted 'Hume's Blush Tea-scented China', which was deemed worthy of a highly official exemption from strict trade embargos.

NEXT PAGE | On the left, 'Hume's Blush Tea-scented China' rose, which Empress Joséphine (right) desperately wanted in her garden

19. Jahrhundert

ROSENPARADIES MALMAISON

KAISERIN JOSÉPHINE UND DIE ROSEN

Napoleon Bonaparte soll seine erste Ehefrau, Joséphine de Beauharnais (1763–1814), leidenschaftlich geliebt haben, ihre große Liebe allerdings galt offenbar – neben ihrem Gatten – den Rosen. Obwohl das französische Kaiserpaar in den Tuilerien in Paris residierte, erwarb Joséphine, mit großzügiger finanzieller Unterstützung Napoleons, im Jahr 1799 Malmaison und ließ das kleine Schloss südlich der großen Stadt nach ihrem Geschmack umgestalten. Die Krönung waren die wunderbaren Gartenanlagen, in der die Blumenliebhaberin seltene Pflanzen aus aller Welt zusammentragen ließ.

Warum es der Kaiserin gerade Rosen besonders angetan hatten, ist nicht belegt. Vielleicht aufgrund ihres dritten Vornamens, Rose? Auf jeden Fall war es Joséphines Ziel, alle bekannten Rosensorten der Welt in Malmaison zu versammeln, koste es, was es wolle! Schon damals waren einige der renommiertesten Rosenzüchter in England beheimatet – mit dem sich Frankreich dummerweise sehr ausdauernd im Krieg befand. Doch Liebe ist, was sie ist – und für Joséphines Rosenleidenschaft ließ Napoleon etwa feindliche Schiffe durchsuchen, sodass viele Pflanzen aus Englands Kolonien ihren Weg nach Malmaison fanden. Das Sahnehäubchen: die begehrte *Hume's Blush Tea-scented China* – sie war sogar eine hochoffizielle Ausnahme von strikten Handelssperren wert.

NÄCHSTE SEITE | Links die Rose *Hume's Blush Tea-scented China*, die Kaiserin Joséphine (rechts) unbedingt in ihrem Garten haben wollte

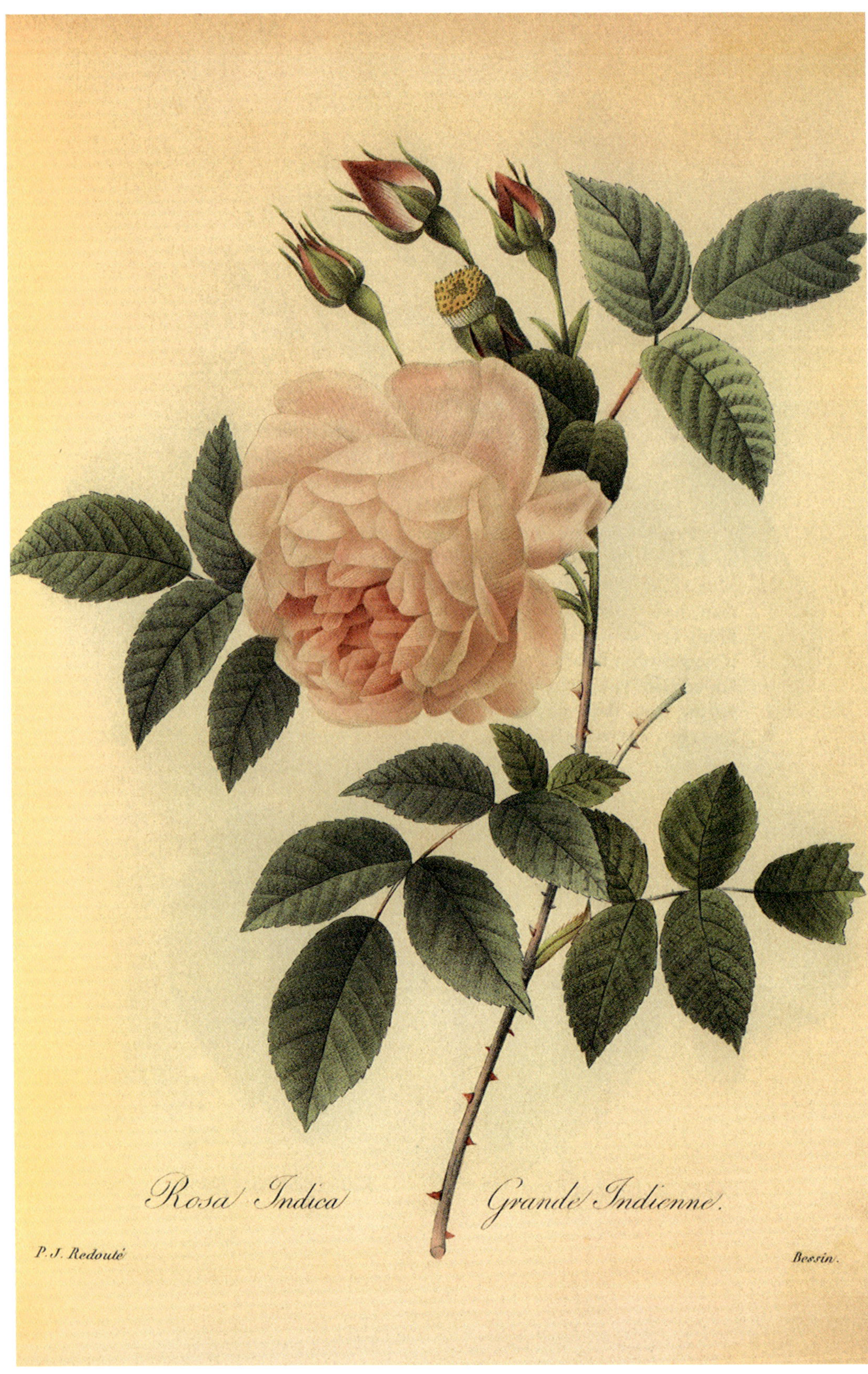

Rosa Indica Grande Indienne.

P. J. Redouté

Bessin.

138

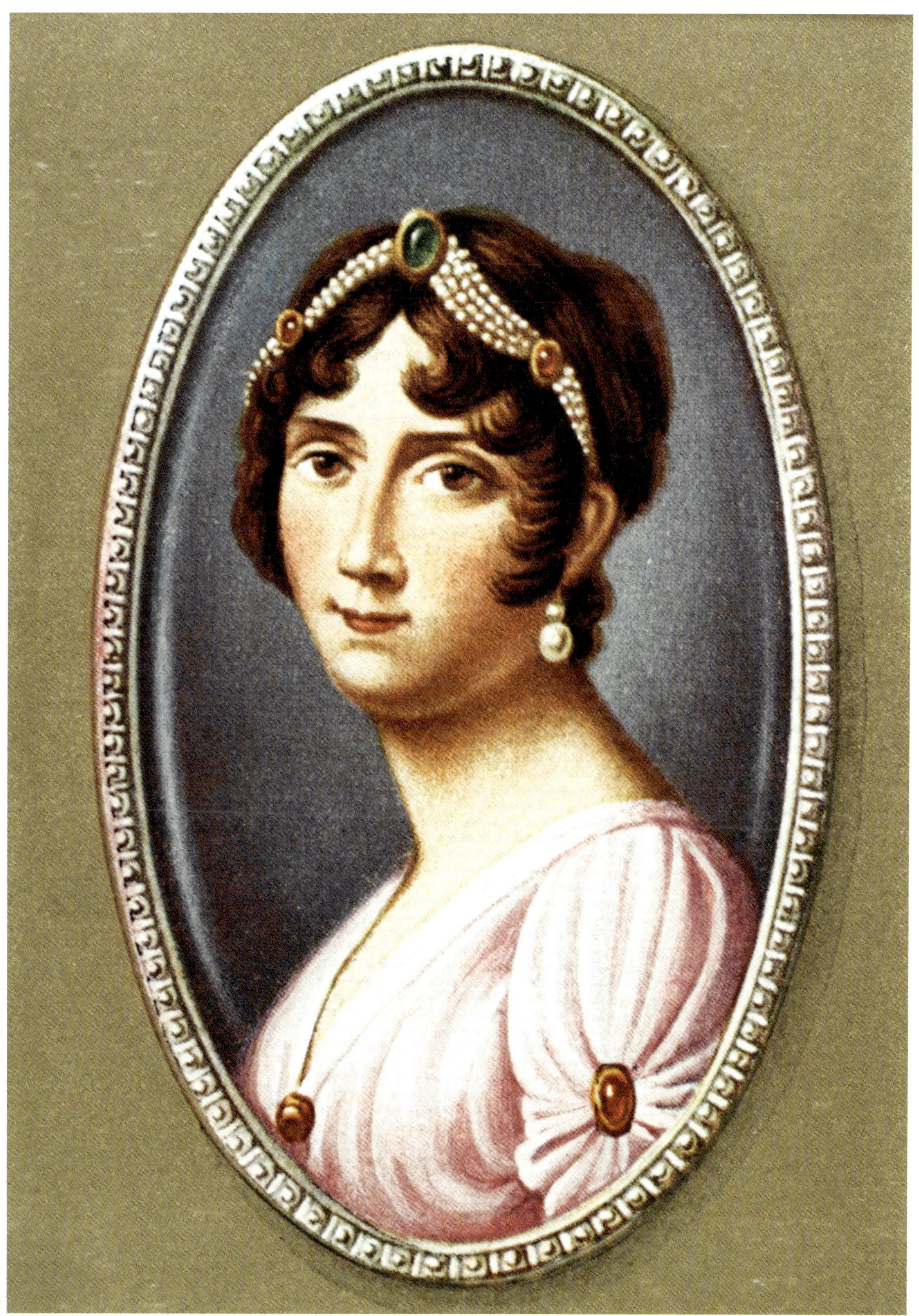

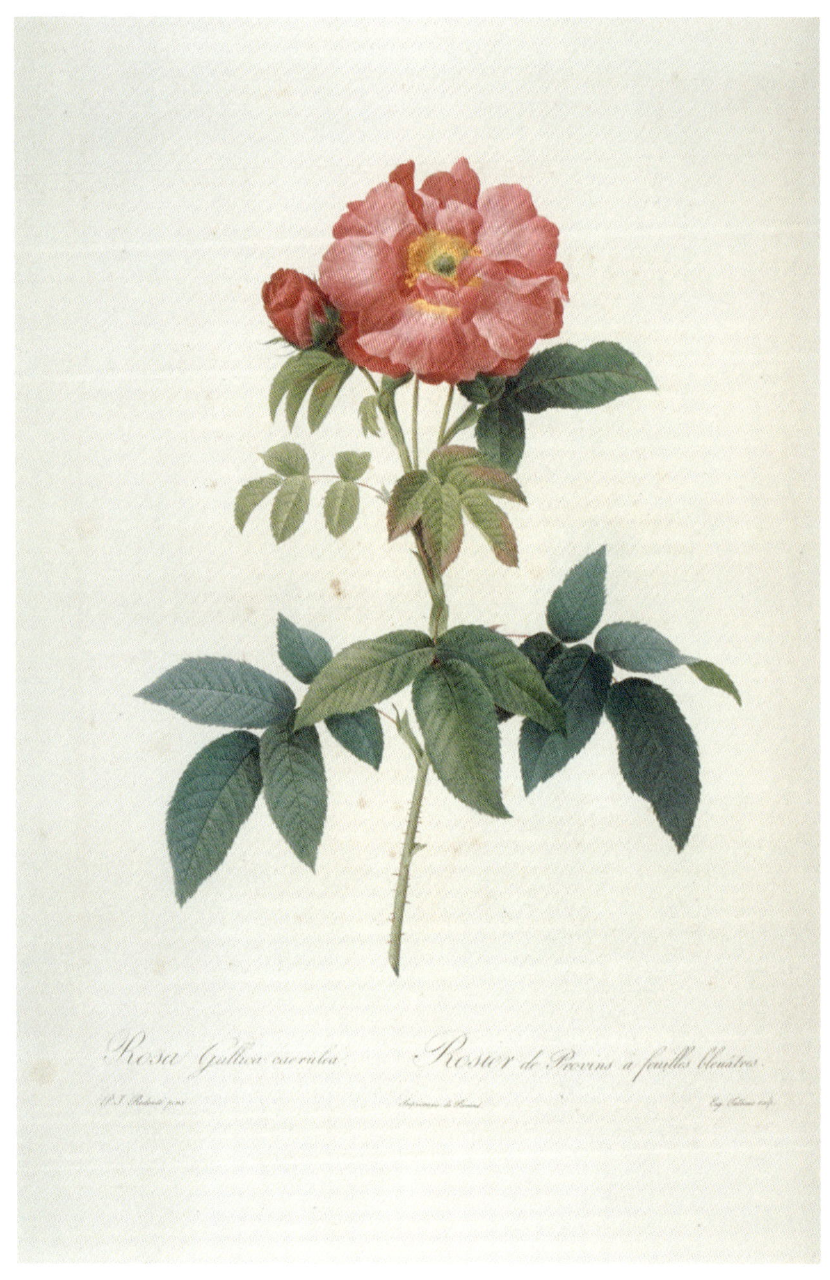

THIS AND THE FOLLOWING PAGES | Pierre-Joseph Redouté's watercolors, exact and aesthetically pleasing, still remain the classic depictions of roses.

DIESE UND FOLGENDE SEITEN | Die ebenso exakten wie ästhetischen Aquarelle von Pierre-Joseph Redouté sind bis heute Klassiker der Rosendarstellungen.

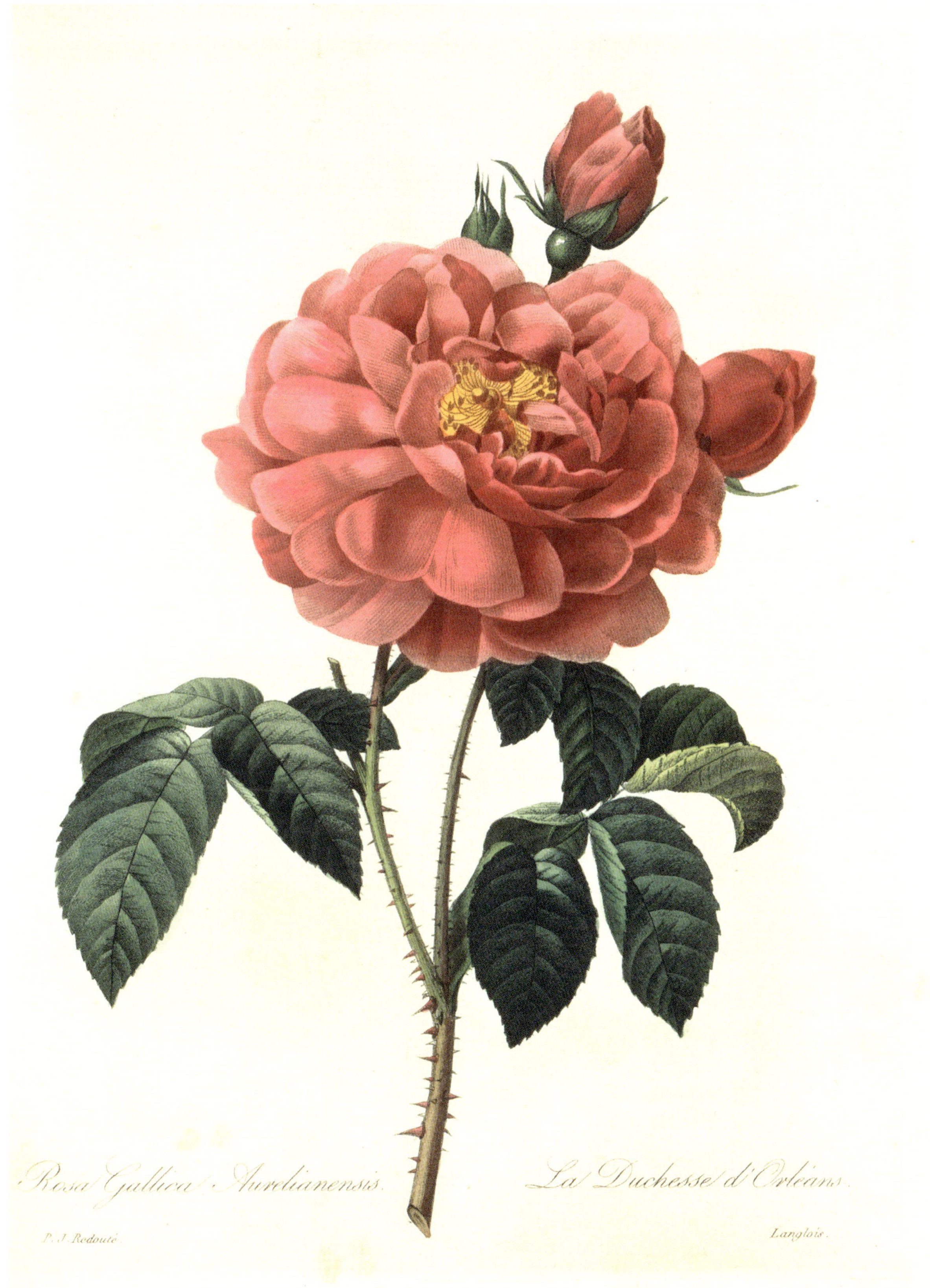

Rosa Gallica Aurelianensis.
La Duchesse d'Orléans.
P. J. Redouté
Langlois.

fMilluptatque quibus ad maximus dollitae

THE RAPHAEL OF THE ROSES

PIERRE-JOSEPH REDOUTÉ'S GRACEFUL AND RICHLY DETAILED WATERCOLORS ARE STILL REPRODUCED BY THE THOUSANDS

The birth of Pierre-Joseph Redouté, in 1759, was perhaps directly into the art world. His father and brothers worked as decorative painters, and by the age of five Pierre-Joseph was demonstrating his extraordinary talent. At thirteen, a child prodigy, he set out to be a portrait painter, but he soon discovered the lush floral still lifes of the Dutch masters, and with that, his love of flowers was awakened. Though he worked as a painter of theatrical scenery in Paris, fulfillment for him lay in drawing the plants of the Jardin du Roi.

His talent did not escape notice; he would later advance to teach drawing at the court of Queen Marie-Antoinette. However, Redouté's art did not truly blossom until Empress Joséphine (p. 136) appointed him to the position of court and flower painter. He was to document her collection at the Jardin de Malmaison. His detailed watercolors, with their nuanced colors, quickly gained fame. Joséphine's protégé published the first of his magnificent volumes on lilies starting in 1802. They were even presented, as an official state gift, to scientists all over the world. However, his three-volume work on Joséphine's favorite flowers, which would endear him to posterity as "Raphael of the Roses", was published only later, between 1817 and 1824. Joséphine had died in May 1814, and so Raphael's patron did not live to see their joint triumph.

Aquarelle und Kupferstiche

DER RAFFAEL DER ROSEN

PIERRE-JOSEPH REDOUTÉS AQUARELLE, SO ANMUTIG WIE DETAILREICH, WERDEN NOCH HEUTE TAUSENDFACH REPRODUZIERT

Vielleicht wurde Pierre-Joseph Redouté bei seiner Geburt im Jahr 1759 die Kunst in die Wiege gelegt: Vater und Brüder waren als Dekorationsmaler tätig, und bereits als Fünfjähriger zeigte Pierre-Joseph seine außerordentliche Begabung. Mit 13 Jahren ging das Wunderkind als Porträtmaler auf Reisen und entdeckte die üppigen Blütenstillleben der niederländischen Meister: Seine Liebe zu Blumen war geweckt! Auch während seiner Zeit als Bühnenmaler in Paris erfüllte ihn das Zeichnen der Pflanzen im Jardin du Roi mehr als seine eigentliche Arbeit.

Sein Talent blieb nicht verborgen, und so avancierte er zum höfischen Zeichenlehrer von Königin Marie Antoinette. Zur wahren Blüte kam Redoutés Kunst jedoch erst, als ihn Kaiserin Joséphine (siehe S. 137) zum Hof- und Blumenmaler ernannte, um ihre Pflanzensammlung im Jardin de la Malmaison zu dokumentieren. Seine detaillierten und farbnuancierten Aquarelle erlangten schnell Berühmtheit. Joséphines Protegé veröffentlichte ab dem Jahr 1802 den ersten seiner Prachtbände über Lilien, die sogar Wissenschaftlern aus aller Welt als offizielles Staatsgeschenk überreicht wurden. Sein dreibändiges Werk über Joséphines Lieblingsblumen, das ihn der Nachwelt als „Raffael der Rosen" bekannt machen sollte, wurde jedoch erst zwischen 1817 und 1824, also nach Joséphines Tod im Mai 1814 veröffentlicht, und so erlebte seine Gönnerin den gemeinsamen Triumph nicht mehr.

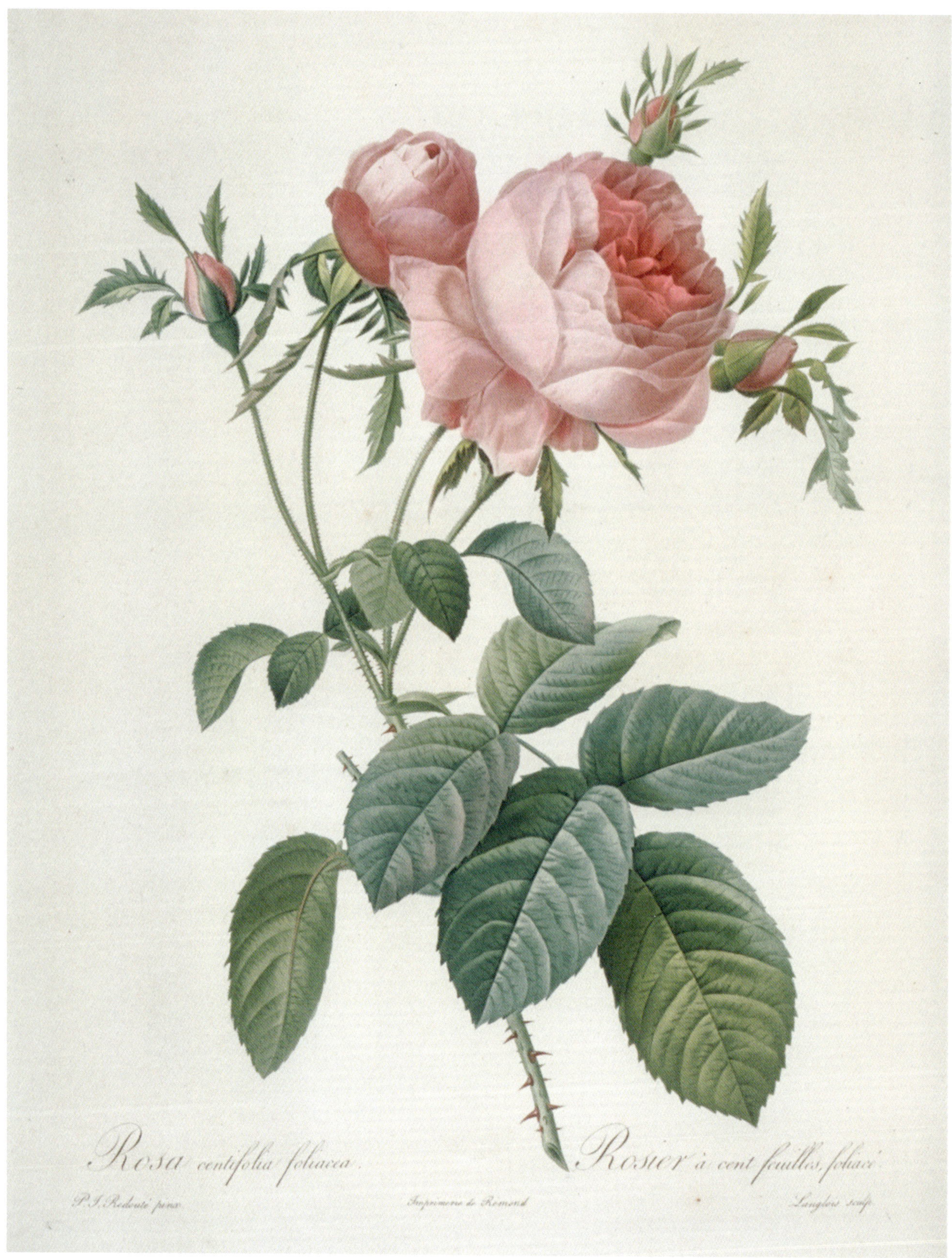

Rosa centifolia foliacea.

Rosier à cent feuilles, foliacé

P.J. Redouté pinx

Imprimerie de Remond

Langlois sculp

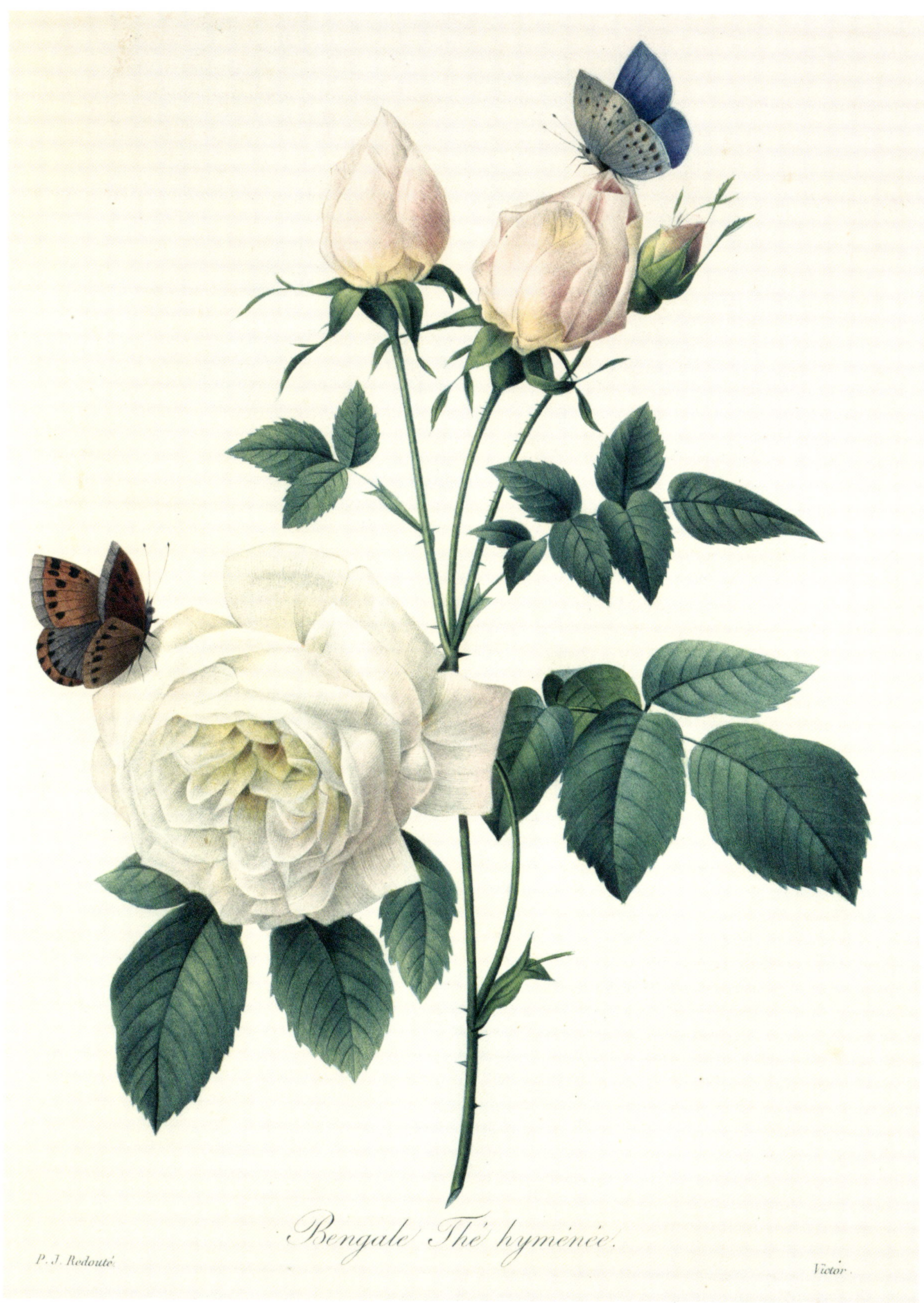

Bengale Thé hyménée.

P. J. Redouté.

Victor.

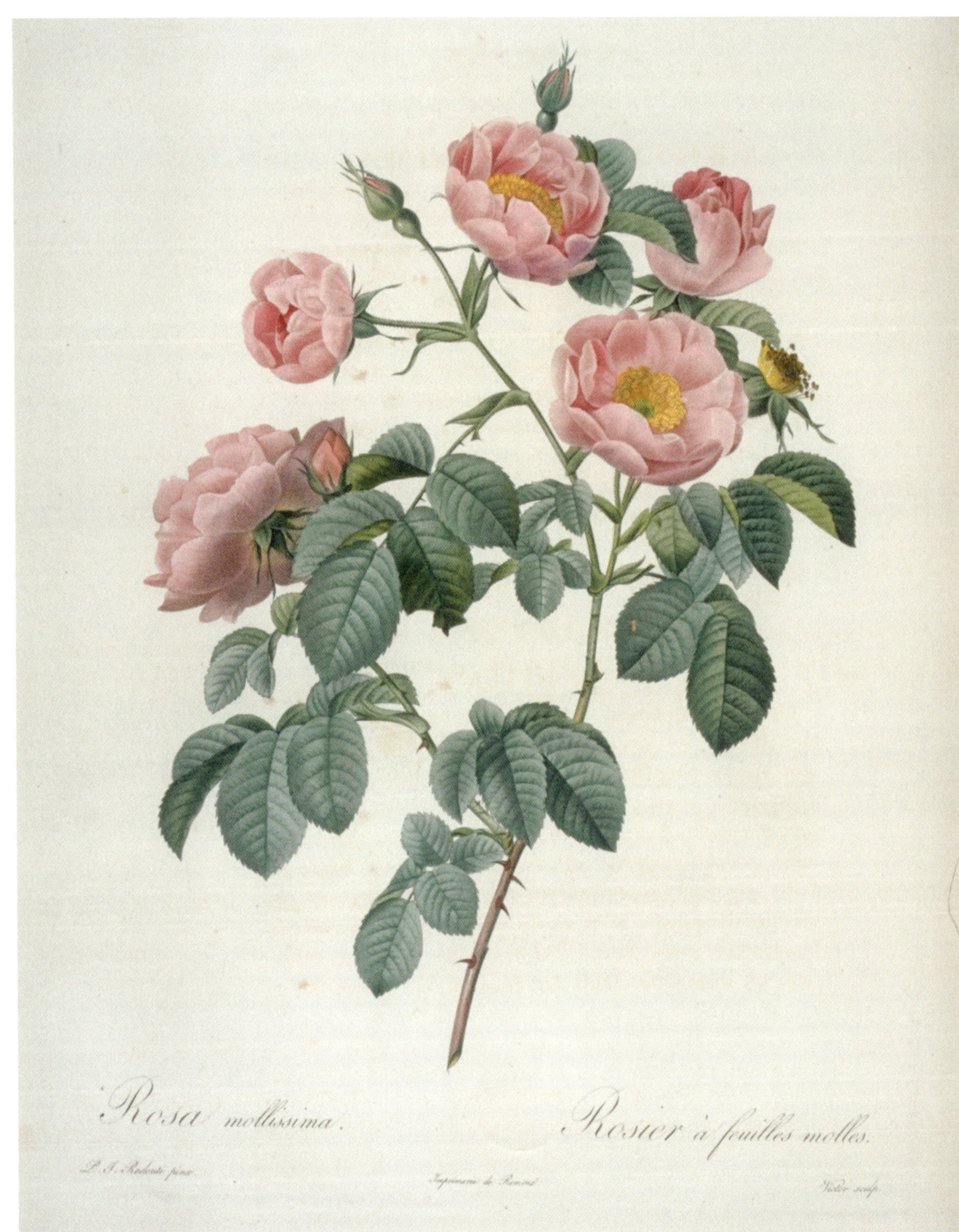

Rosa mollissima.

Rosier à feuilles molles.

P. J. Redouté pinx.

Imprimerie de Remond.

Victor sculp.

Rosa Damascena.
Rosier de Cels.
P.J. Redouté pinx.
Imprimerie de Remond
Charlin sculp.

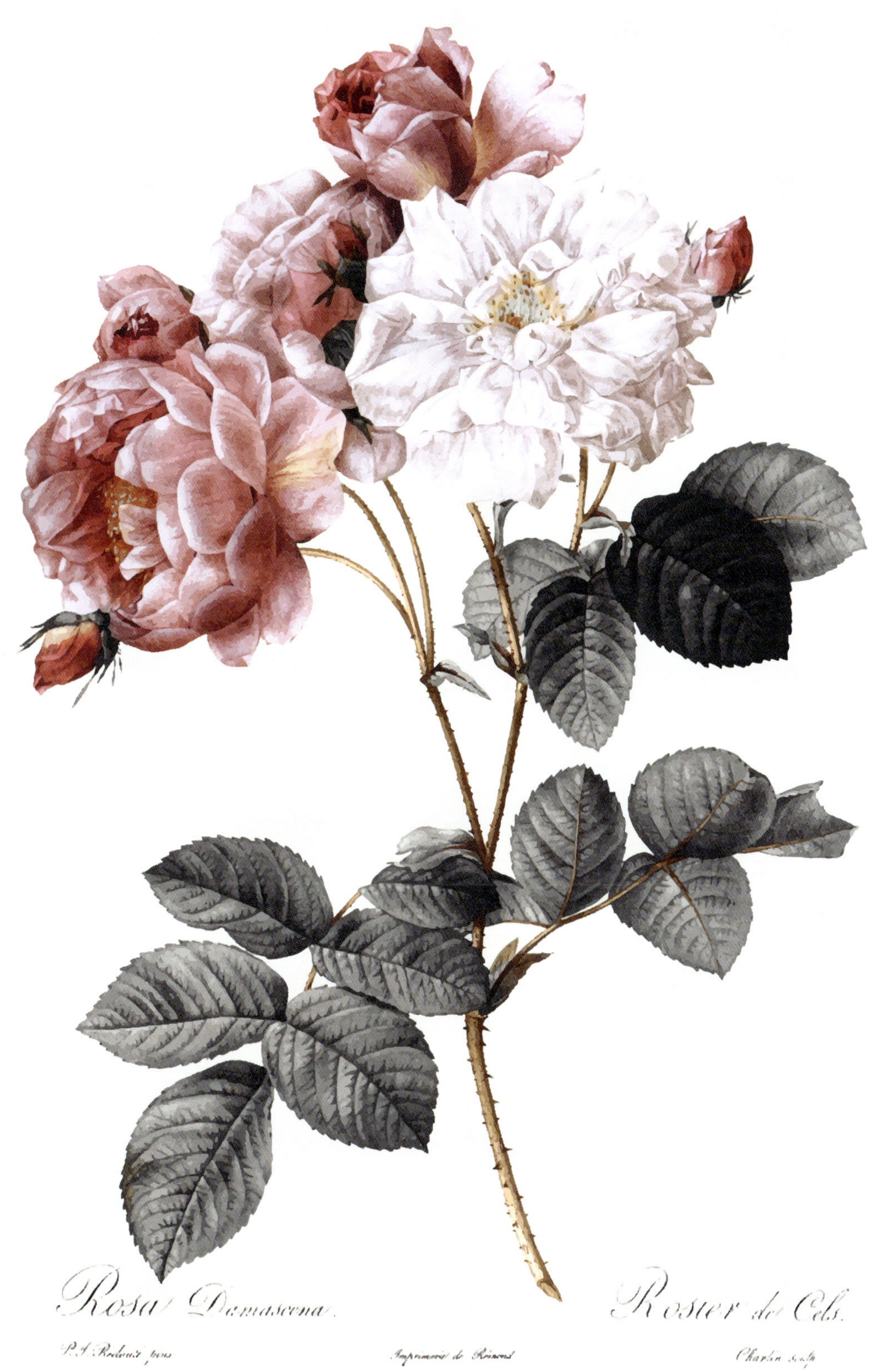

147

ELIZABETH'S MIRACLE OF THE ROSES

TWO SAINTS AND THE LEGEND OF THE ROSE

Saint Elizabeth of Hungary is revered as the epitome of charity. Throughout her life, the Hungarian princess and Thuringian landgravine was renowned for her commitment to the poor and the sick. Even today, she is associated with the legendary miracle of the roses.

At a young age, the pious Elizabeth renounced the pompous lifestyle of princes and preferred to devote herself to the sick, weak, and poor—much to the chagrin of her noble family and of her husband Ludwig, who allegedly forbade this involvement of hers, under threat of punishment. Defying him, she had endeavored to take a basket of bread to the needy; but Ludwig intercepted her, and he forced her to reveal what she had hidden in the basket. Bravely, firm in her faith, she replies, "Roses, my lord"—whereupon the enraged husband tears the cloth from Elizabeth's basket to reveal not loaves of bread but roses.

Whether Elizabeth herself was saved from her husband's wrath by the miracle of the roses is questionable. Contemporary sources describe her marriage to Louis of Thuringia as a happy and harmonious one, noting that he expressly supported his wife's charitable efforts. After the death of her husband, however, Elizabeth's concern for the poor did not sit well with her husband's family. And this perhaps furnishes an explanation for how a legend told about Elizabeth of Portugal, her grandniece and namesake, was posthumously attributed to her, the patron saint of Thuringia. The biographies of the two Elizabeths are indeed similar: as children, both were married to a nobleman; both devoted their lives to piety and asceticism; and in times of famine, both opened the princes' granaries to their poorest subjects. Unlike Louis, however, Elizabeth of

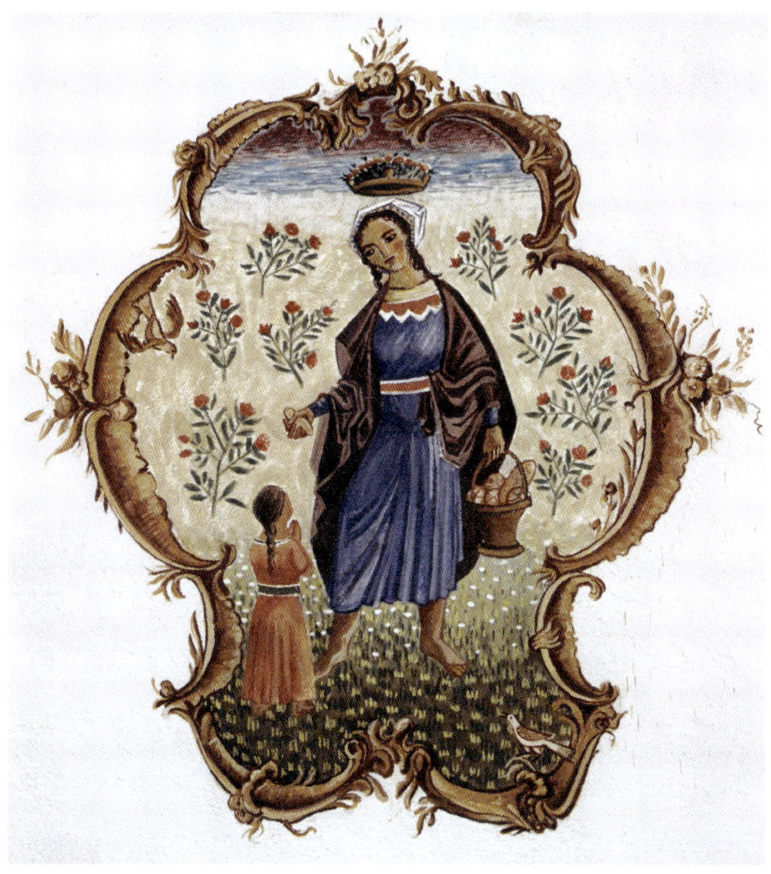

A representation of St. Elizabeth handing a piece of bread to a girl. No roses in her basket—yet.

Darstellung der Heiligen Elisabeth – hier noch ohne Rosen im Korb. Sie gibt einem Mädchen ein Stück Brot.

Portugal's husband, the king Denis, is said not to have approved of her charity, and so her miracle of the roses merges in tradition with the biography of her Thuringian great-aunt.

Eine Wohltäterin wird beschützt

DAS ROSENWUNDER DER ELISABETH

ZWEI HEILIGE UND EINE LEGENDE

Als Inbegriff der Nächstenliebe wird die heilige Elisabeth von Thüringen verehrt: Schon zeitlebens war die ungarische Prinzessin und thüringische Landgräfin weithin für ihren Einsatz für Arme und Kranke bekannt. Noch heute verbindet man mit ihr die Legende vom Rosenwunder.

Die fromme Elisabeth entsagt schon in jungen Jahren dem prunkvoll-fürstlichen Lebensstil und widmet sich lieber Kranken, Schwachen und Armen – sehr zum Missfallen ihrer adligen Familie und ihres Ehemanns Ludwig, der ihr Engagement angeblich bei Strafe untersagt. Als sie sich ihm widersetzt und einen Korb voll Brot zu den Bedürftigen bringen will, fängt Ludwig Elisabeth ab und zwingt sie zu offenbaren, was sich im zugedeckten Korb befindet. Tapfer und in festem Glauben entgegnet sie „Rosen, mein Herr" – und als der erzürnte Gatte das Tuch von Elisabeths Korb reißt, finden sich darin tatsächlich statt der Brote Rosen!

Ob es nun wirklich Elisabeth selbst war, die durch das Rosenwunder vor dem Zorn ihres Ehemanns bewahrt wurde, ist jedoch fraglich. Zeitgenössische Quellen beschreiben ihre Ehe mit Ludwig von Thüringen als glücklich und harmonisch und vermerken, dass dieser den mildtätigen Einsatz seiner Ehefrau ausdrücklich unterstützte. Nach dem Tod ihres Gatten hingegen kam Elisabeths Fürsorge für die Armen bei der Familie ihres Mannes nicht gut an. Vielleicht erklärt sich so, dass eine Legende, die man später über ihre Großnichte und Namenserbin Elisabeth von Portugal erzählte, posthum der thüringischen Landespatronin zugeschrieben wurde. Die Biografien der beiden Elisabeths ähneln sich

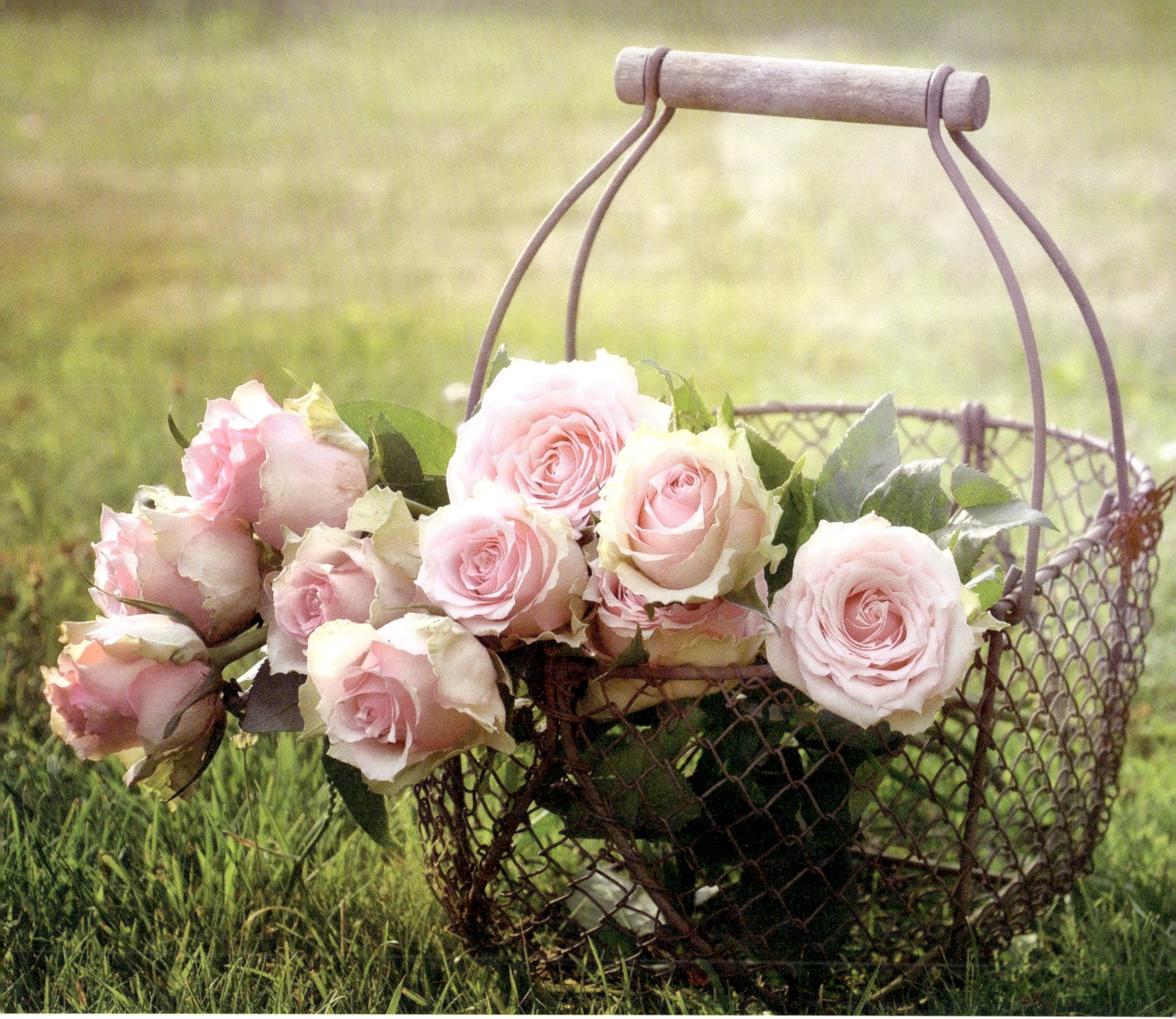

zudem: Beide wurden schon in Kindertagen mit einem adligen Gatten vermählt, waren einem Leben in Frömmigkeit und Askese zugetan und öffneten zu Hungerszeiten die fürstlichen Kornkammern für die Ärmsten unter ihren Untertanen. Im Gegensatz zu Ludwig soll jedoch Elisabeth von Portugals Gatte König Dinis ihre Mildtätigkeit nicht gutgeheißen haben, und so verschmilzt ihr Rosenwunder in der Überlieferung mit der Biografie ihrer thüringischen Großtante.

Love of roses in gardens and poetry

ROSE, OH PURE CONTRADICTION

ENIGMATIC ROSES ARE RAINER MARIA RILKE'S FINAL POETIC LEGACY

Nature looms large in Rainer Maria Rilke's poetry; flowers, especially roses, recur throughout his lyrical work. An anecdote from his time in Paris holds that there was a beggar woman who would sit in the same spot, lethargically, day in and day out, and that he once gave her a rose instead of money or food, in the belief that her soul also must be nourished. When Rilke settled in Switzerland to live after a life of restless drifting and wandering, he devoted himself not only to poetry but also to his garden, to his beloved flowers. His favorites are said to have been the rose varieties 'La France' and 'Mrs John Laing', with their intense pink colors and fine fragrance.

Rilke willed that these lines should be immortalized on his tombstone in the small mountain cemetery of Raron, in Valais: *Rose, oh reiner Widerspruch, Lust / Niemandes Schlaf zu sein / unter soviel / Lidern* (Rose, oh pure contradiction, desire / to be no-one's sleep beneath / so many lids). A glance at the diaries of Rilke's patroness, Princess Marie von Thurn und Taxis, perhaps solves the riddle posed by these lines: With the words, "Oh Rainer! I object to that!", she is thought to have found flaw with an excessively floral quality in Rilke's poems.

NEXT PAGE | The 'Mrs. John Laing' variety is said to have been one of Rilke's favorite roses. The poet lived in Valais for five years before his death, in 1926, surrounded by his beloved garden.

Blumenliebe in Garten und Dichtung

„ROSE, OH REINER WIDERSPRUCH"

RÄTSELHAFTE ROSEN SIND RAINER MARIA RILKES LETZTES POETISCHES VERMÄCHTNIS

In Rainer Maria Rilkes Dichtung spielt die Natur eine große Rolle, Blumen und vor allem Rosen sind wiederkehrende Themen in seinem lyrischen Werk. Aus seiner Zeit in Paris ist eine Anekdote überliefert, in der er einer Bettlerin, die tagein, tagaus lethargisch am selben Platz saß, anstatt von Geld oder Essen eine Rose überreichte – ihm war es wichtig, auch ihrer Seele Nahrung zu geben. Als Rilke nach einem Leben voll unruhigem Getriebensein und Wanderschaft in seinem letzten Lebensabschnitt in der Schweiz Heimat fand, widmete er sich nicht nur in der Dichtung, sondern auch im Garten seinen Lieblingsblumen. Die Rosensorten *La France* und *Mrs John Laing* mit ihrem intensiven Rosaton und feinen Duft sollen seine Lieblinge gewesen sein.

Schon zu Lebzeiten verfügte Rilke, dass auf seinem Grabstein auf dem kleinen Bergfriedhof von Raron im Kanton Wallis folgende Zeilen verewigt werden sollten: „Rose, oh reiner Widerspruch,/ Lust,/ Niemandes Schlaf zu sein/ unter soviel /Lidern." Der Sinn der rätselhaften Zeilen erschließt sich vielleicht durch einen Blick in die Tagebücher von Rilkes Mäzenin Fürstin Marie von Thurn und Taxis, die ein Zuviel an Blumigkeit in Rilkes Gedichten bemängelt haben soll mit den Worten: „Oh Rainer! Widerspruch lege ich dagegen ein!"

NÄCHSTE SEITE | Die Sorte *Mrs. John Laing* soll zu Rilkes Lieblingsrosen gehört haben. Der Dichter lebte vor seinem Tod 1926 fünf Jahre im Wallis, umgeben von seinem geliebten Garten.

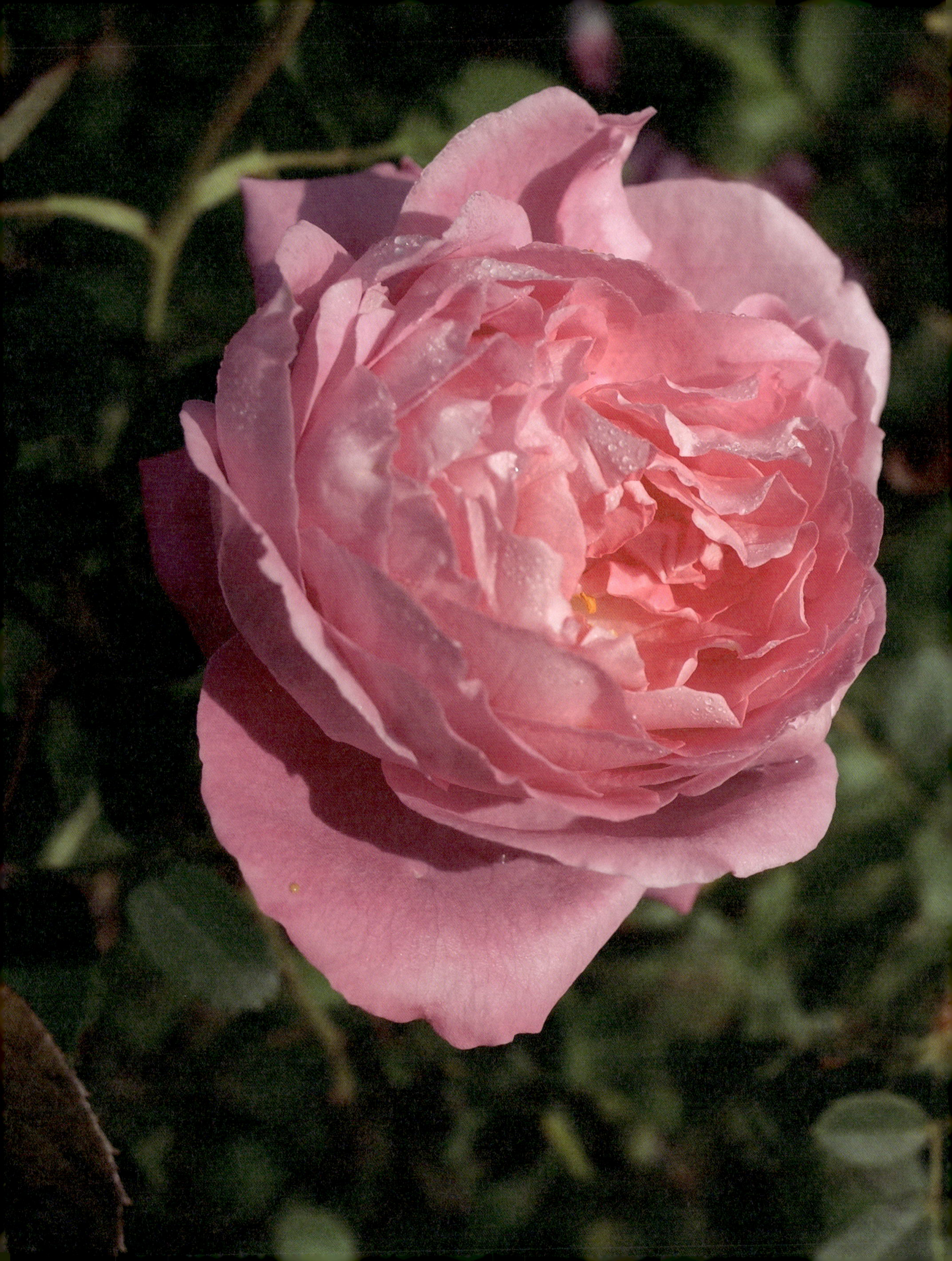

Twentieth century

GLORIA DEI

A ROSE OF PEACE, WITH MANY NAMES

In 1945, a delicate yellow rose was supposed to instill, in all the world, the idea of peace. But this rose's history as a symbol of international solidarity had begun earlier, in 1939, when Francis Meilland, a Frenchman, sent his "most beautiful rose in the world" to Germany, Italy, the United States, and Great Britain, to unite rose-lovers around the globe through this new variety. The war abruptly interrupted all global cooperation, and so the new rose received a new name in each country. For France, Meilland dedicates it to his mother with the name 'Madame A. Meilland'; in Italy, it is called 'Gioia', reaching for joy in times of darkness. Pfizer, the German rose-breeder, anoints it 'Gloria Dei' (to the glory of God), a protest against the un-Christian acts of the Nazis. In the United States, in April 1945, with people longing for peace as much as anyone in the war-torn world, the breeder Pyle called it 'Peace' as a sign of hope.

Later, when the Charter of the United Nations was signed in San Francisco (setting a seal, as it were, on world peace), each delegate would find a rose in their hotel room, and with it an inspiring message: "Here is the Rose 'Peace', baptized in Pasadena, California, on the day Berlin fell. We hope the 'Peace' rose will influence men's thoughts for everlasting world peace".

RIGHT AND NEXT PAGE | 'Gloria Dei' is the most famous and best-selling yellow hybrid tea in the world.

RECHTE UND NÄCHSTE SEITE | Die *Gloria Dei* ist die berühmteste und meistverkaufte gelbe Teehybride der Welt.

20. Jahrhundert

GLORIA DEI

FRIEDENSROSE MIT VIELEN NAMEN

Eine Rose in zartem Gelb sollte 1945 den Friedensgedanken in die Welt tragen, ihre Geschichte als Symbol internationaler Verbundenheit beginnt jedoch schon 1939: Der Franzose Francis Meilland schickt die „schönste Rose der Welt" nach Deutschland, Italien, in die USA und nach Großbritannien, um mit seiner Neuzüchtung Rosenfreunde rund um den Erdball zu vereinen. Der Krieg unterbricht jäh die globale Zusammenarbeit, und so erhält die neue Rose in jedem Land einen anderen Namen: In Frankreich widmet Meilland sie mit *Madame A. Meilland* seiner Mutter, in Italien wird sie *Gioia* genannt, auf der Suche nach Freude in düsteren Zeiten. Der deutsche Rosenzüchter Pfizer tauft sie zum Ruhm Gottes *Gloria Dei*, als Zeichen gegen das unchristliche Treiben der Nazis. Und in Amerika? Sehnt man sich genauso nach Frieden wie überall auf der kriegsgebeutelten Welt – US-Züchter Pyle nennt die Rose im April 1945 hoffnungsfroh *Peace*.

Als kurz darauf in San Francisco die Unterzeichnung der UN-Charta den Weltfrieden besiegeln soll, findet jeder Delegierte im Hotelzimmer eine Rose mit inspirierender Botschaft: „Hier ist die Rose *Peace*, die in Pasadena in Kalifornien an dem Tag, an dem Berlin fiel, getauft wurde. Möge diese Rose alle Menschen guten Willens beeinflussen, für die Schaffung eines gerechten und dauerhaften Friedens."

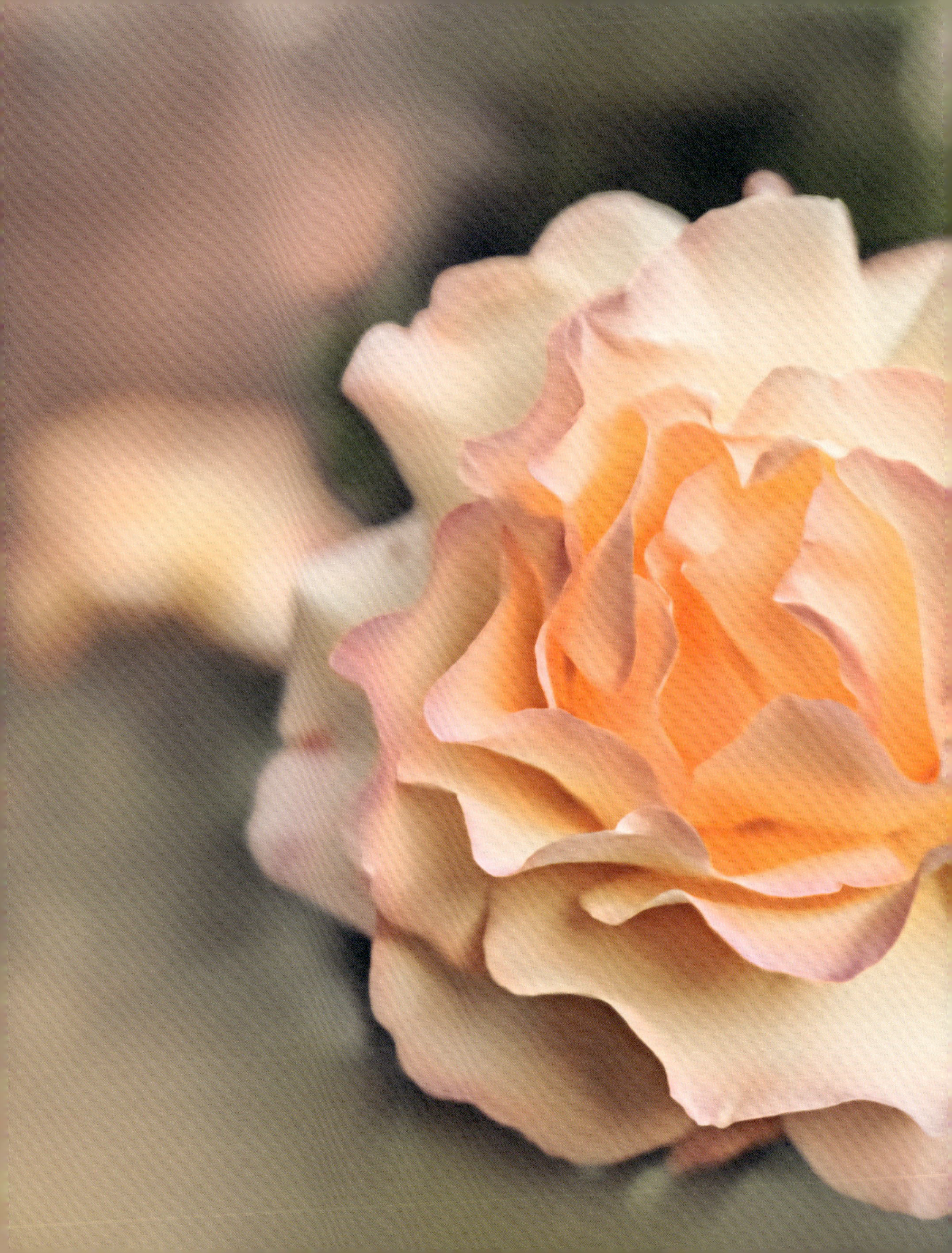

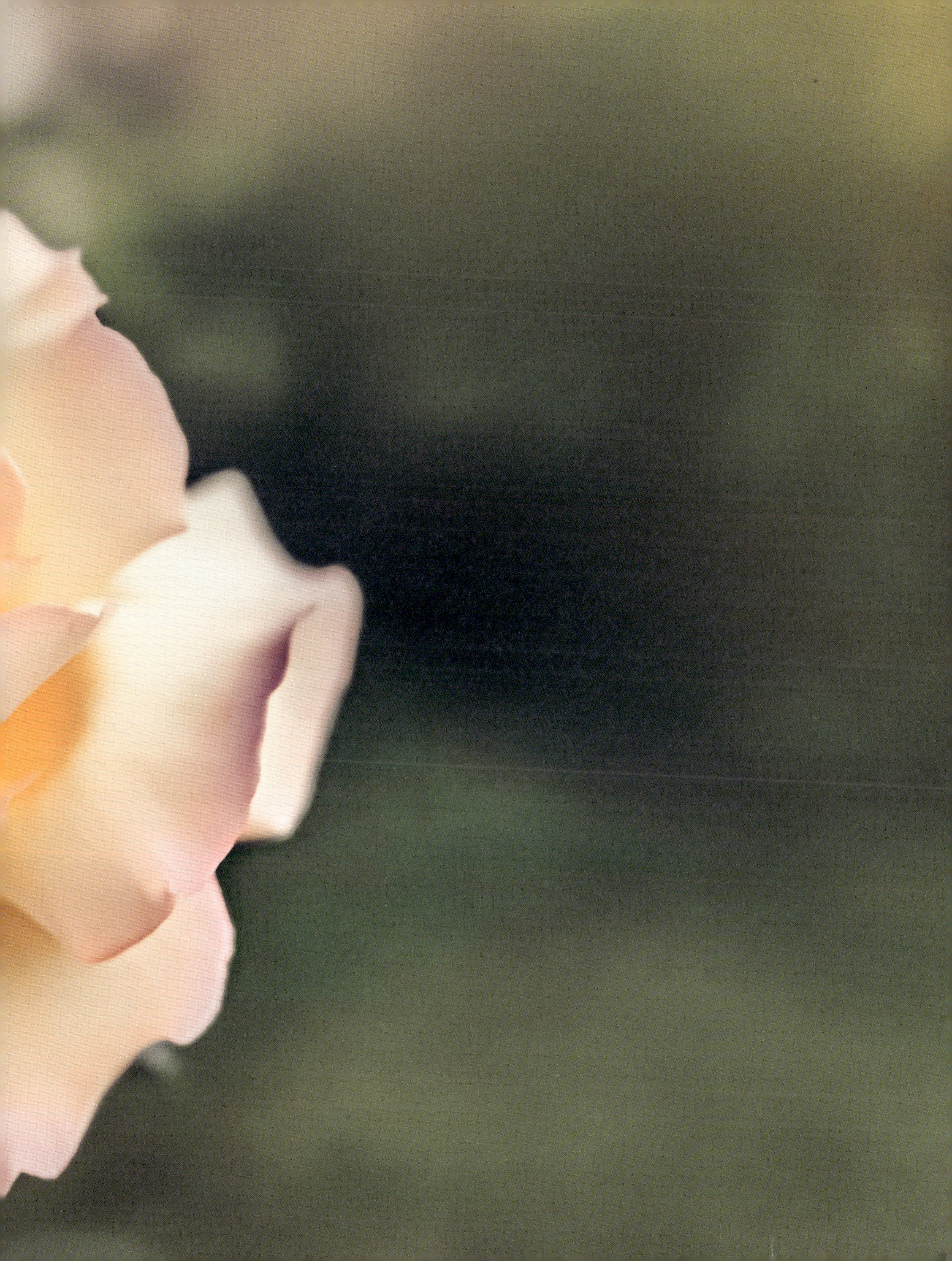

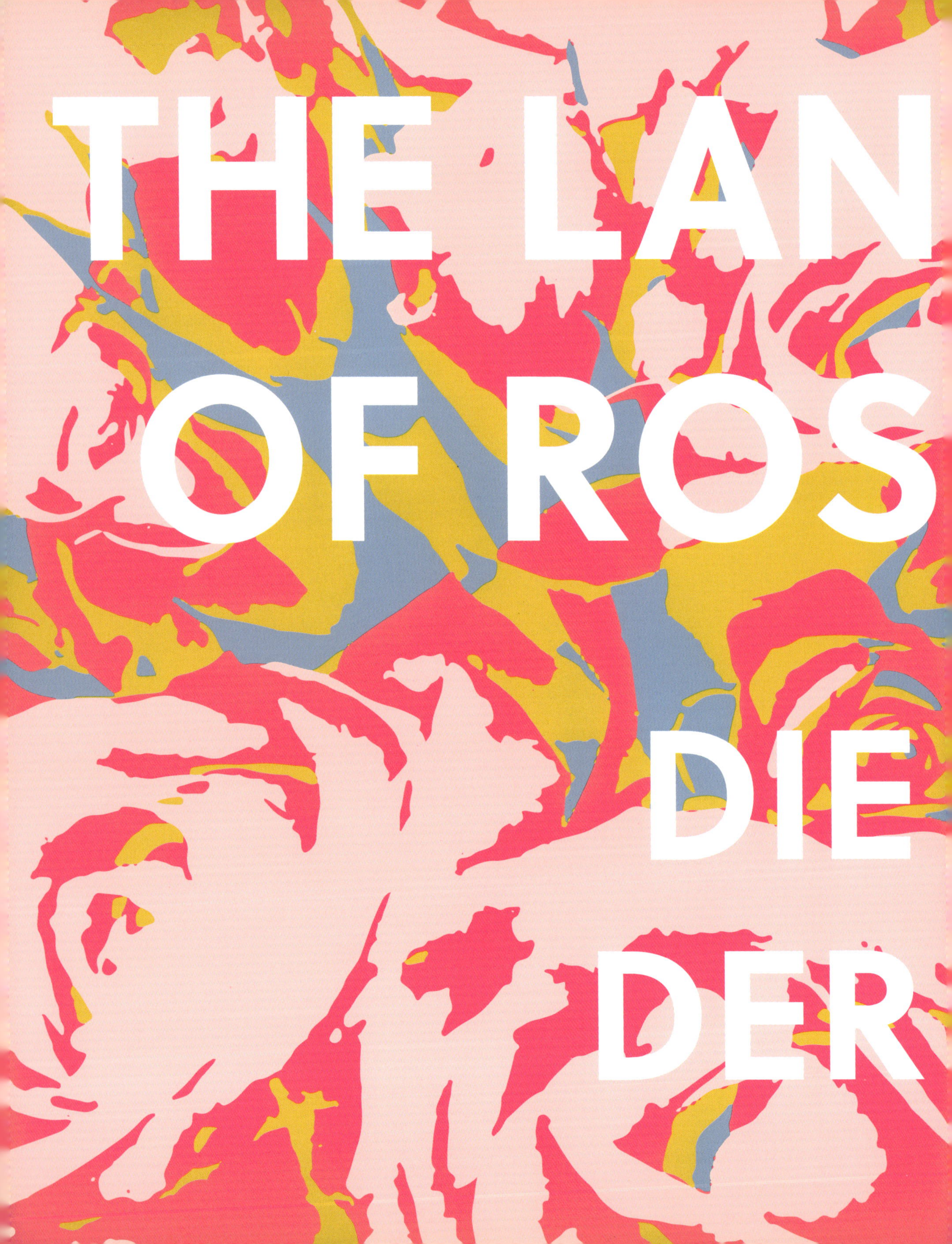
THE LAN
OF ROS
DIE
DER

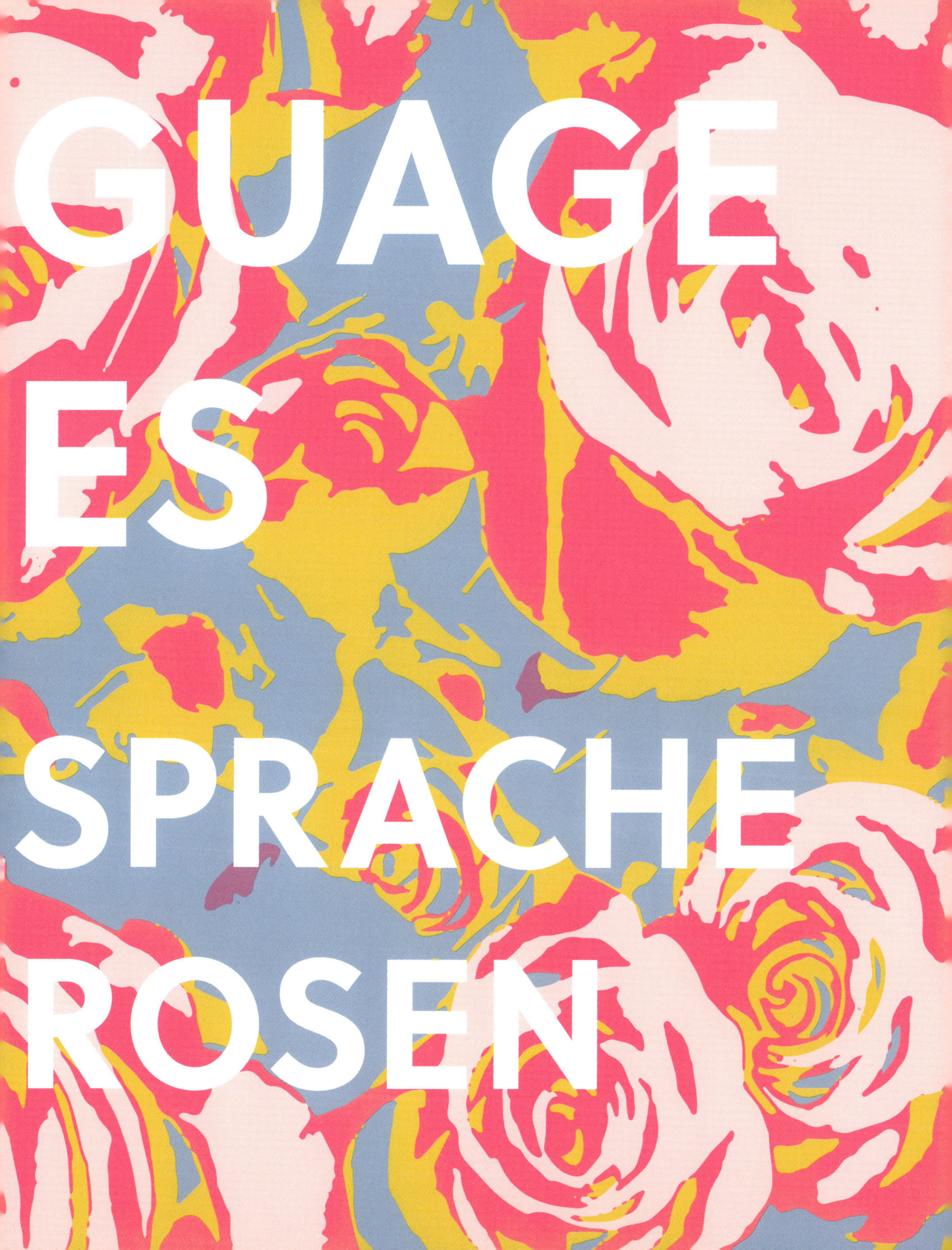
GUAGE
ES
SPRACHE
ROSEN

THE LANGUAGE OF ROSES

INFATUATION OR LOVE ETERNAL? JEALOUSY OR REGRET? EMOTIONS FLOURISH IN THE LANGUAGE OF FLOWERS

A discrete message can be communicated *sub rosa* (under the rose). As you can probably imagine, once upon a time it was mainly lovers sending greetings through flowers, with an unspoken message that would have been unbecoming if uttered openly and clearly in words. In the Ottoman Empire, there was even a language of flowers known as *selam*, to which Europeans were introduced at the beginning of the eighteenth century by the English writer Lady Mary Wortley Montagu. Living for a time in Istanbul, she was fascinated by the elaborate code of flowers through which highly nuanced messages could be exchanged, wordlessly, among the otherwise strictly secluded private chambers of high-society women. It may have been in the Victorian era, as social conventions progressively squeezed interpersonal relationships into an ever-tightening corset, that the language of flowers attained its greatest popularity. Not only did individual species of flowers and the colors of blossoms convey

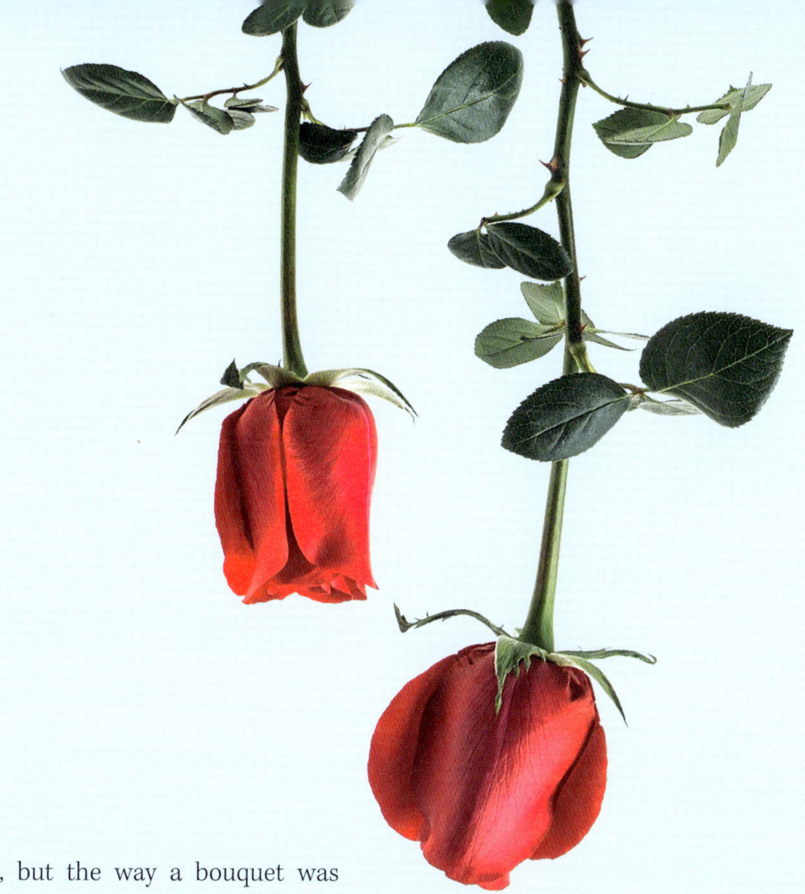

a coded message, but the way a bouquet was bound and decorated, or even the way it was presented, carried hidden meaning.

A message shrouded in flowers could also be conveyed through the number of blossoms sent, and this was particularly true of bouquets of roses. To this day, each number of roses sent as a gift is thought to have its own meaning, and it goes far beyond the convention of only ever arranging an odd number of roses in a bouquet. Everything begins with a single rose—yes, it is love at first sight. Two roses signal a belief that the love is mutual. If this human object of the heart's fancy answers in kind—with two roses—the deal is sealed, and three red roses will now read like the magic words, "I love you." Six roses disregard the rule of odd numbers: the half-dozen signals readiness to go all-in; it says, "let me be thine." While it is unclear whether it takes five or ten roses to say, "you are perfect," will anyone seriously doubt the message? With nine red roses, the message is eternal love. If these turn into ninety-nine roses, the statement becomes even more emphatic: "I will love you as long as I live." But please count with care: if you receive 108 red roses, it might be a marriage proposal. And if you present a bouquet of fifty roses after many years of a long relationship, the message is clear: it is love, filled with gratitude and free of regret.

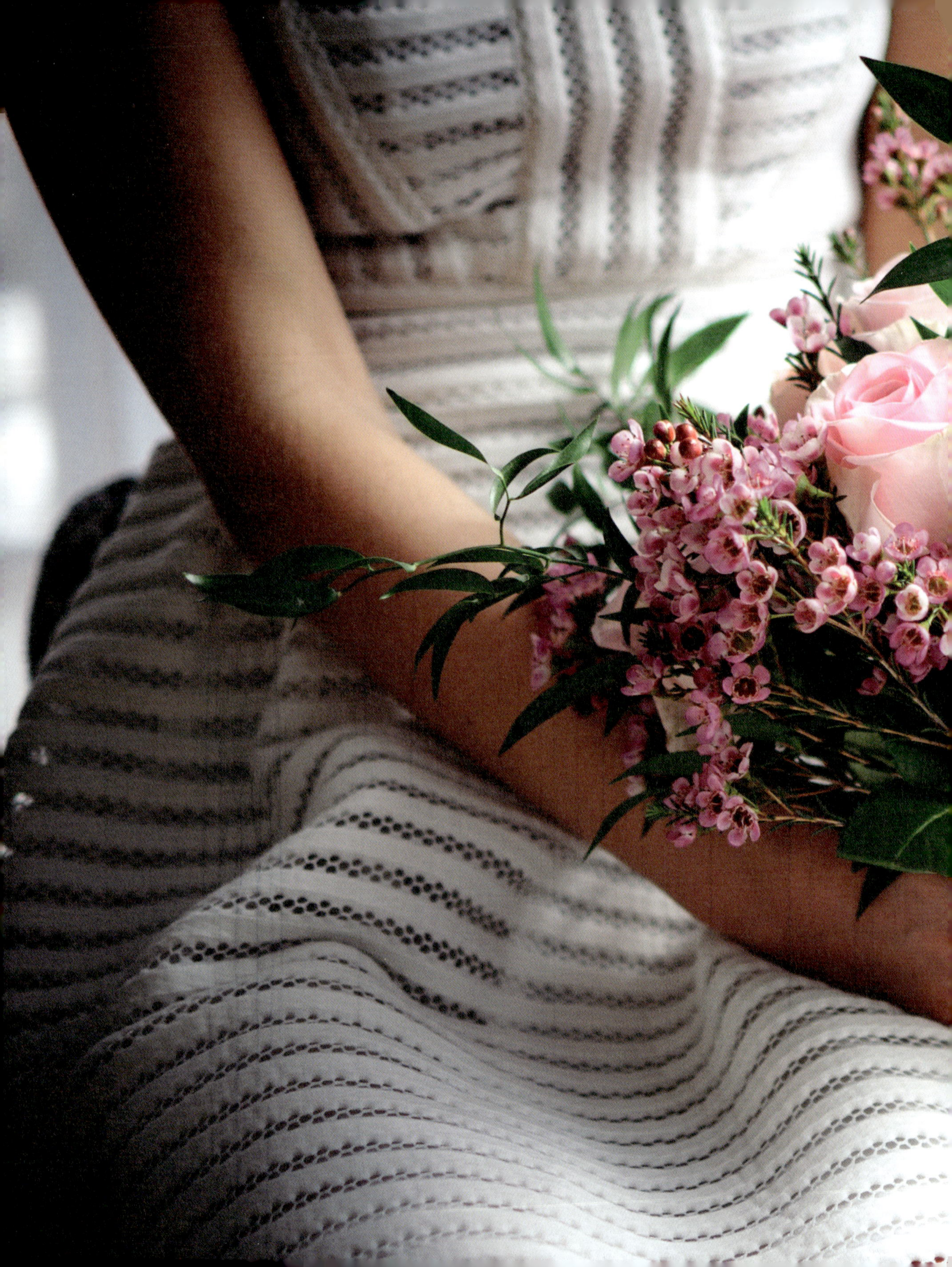

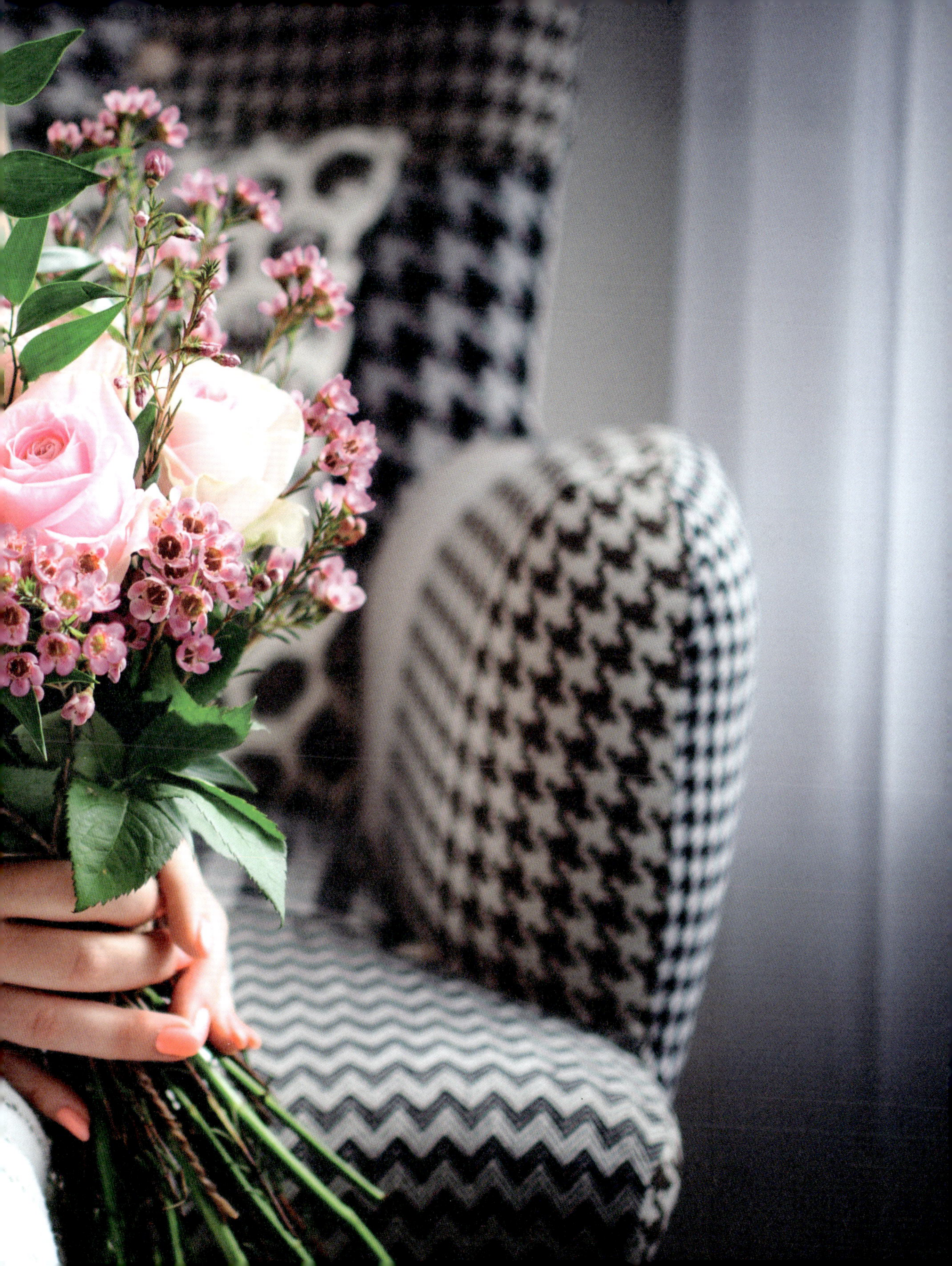

DIE SPRACHE DER ROSEN

VERLIEBTHEIT ODER EWIGE LIEBE? EIFERSUCHT ODER REUE? DIE SPRACHE DER BLUMEN LÄSST EMOTIONEN AUFBLÜHEN

„Durch die Blume gesagt" – so nennen wir es, wenn eine Botschaft subtil vermittelt werden soll. Einst waren es meist stille Nachrichten unter Liebenden, die mit Blumengrüßen ausgedrückt wurden – dies in offenen, klaren Worten zu kommunizieren hätte sich in vergangenen Tagen wohl kaum geziemt. Im Osmanischen Reich existierte sogar eine eigene Sprache der Blumen, *Selam* genannt, die über die britische Schriftstellerin Lady Mary Wortley Montagu Anfang des 18. Jahrhunderts in Europa bekannt wurde: Sie selbst hatte einige Zeit in Istanbul gelebt und war fasziniert von den ausgefeilten Blumencodes, mit denen fein nuanciert wortlose Botschaften unter den ansonsten streng abgeschiedenen privaten Gemächern vornehmer Damen ausgetauscht werden konnten. Zu viktorianischen Zeiten, als die gesellschaftlichen Konventionen das zwischenmenschliche Miteinander in ein immer engeres Korsett schnürten, erlangte die Sprache der Blumen ihre vielleicht größte Popularität. Nicht nur die einzelnen Blumenarten und Blütenfarben vermittelten eine codierte Bedeutung: Auch wie ein Strauß gebunden und ausgeschmückt war und sogar die Art und Weise, wie das Bouquet überbracht wurde, trugen eine versteckte Botschaft in sich.

Blumig verhüllte Botschaften ließen sich zudem über die Anzahl der Blumen, speziell in Rosenbouquets, übermitteln. Auch heute noch wird der Anzahl von Rosen, die man verschenkt, eine bestimmte Bedeutung zugesprochen – und diese geht deutlich über die Konvention hinaus, Rosen nur in ungerader Anzahl in Bouquets zu arrangieren. Alles beginnt mit einer einzelnen Rose – ja, es ist Liebe auf den ersten Blick.

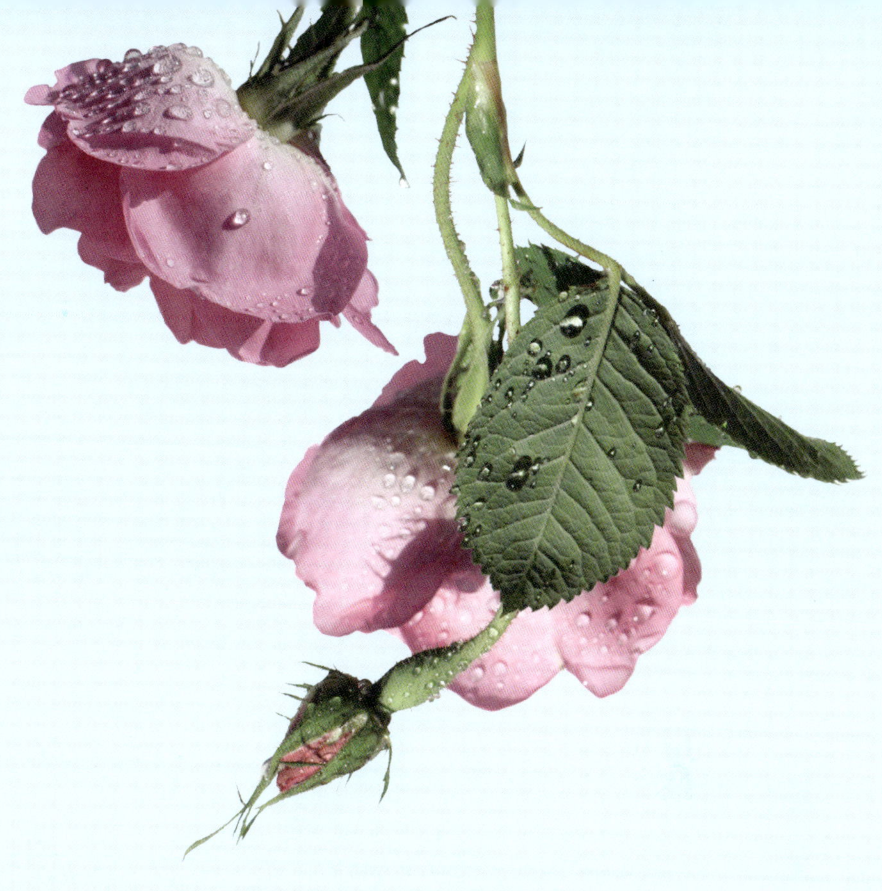

Zwei Rosen signalisieren den Glauben an die Gegenseitigkeit der Liebe. Wenn der Herzensmensch ebenfalls mit zwei Rosen antwortet, steht dem Glück nichts mehr im Weg: Es ist Zeit für drei rote Rosen, für die magischen drei Worte: „Ich liebe dich." Mit sechs roten Rosen setzt man sich über die Regel der ungeraden Zahlen hinweg – das halbe Dutzend signalisiert die Bereitschaft, sich ganz hingeben zu wollen: „Lass mich dein sein!" Ob es nun fünf oder zehn Rosen sind, die sagen „Du bist perfekt", ist unklar – aber wer wird bei solch einer Botschaft schon zweifeln? Mit neun roten Rosen erklärt man die ewige Liebe – und werden daraus sogar 99 Rosen, ist die Botschaft noch eindringlicher: „Ich werde dich mein Leben lang lieben!" Aber bitte genau sein beim Zählen: Erhält man 108 rote Rosen, könnte es ein Heiratsantrag sein. Und wer nach vielen Beziehungsjahren ein Bouquet von 50 Rosen überreicht, zeigt ganz deutlich: Es ist Liebe voll Dankbarkeit und ohne Reue!

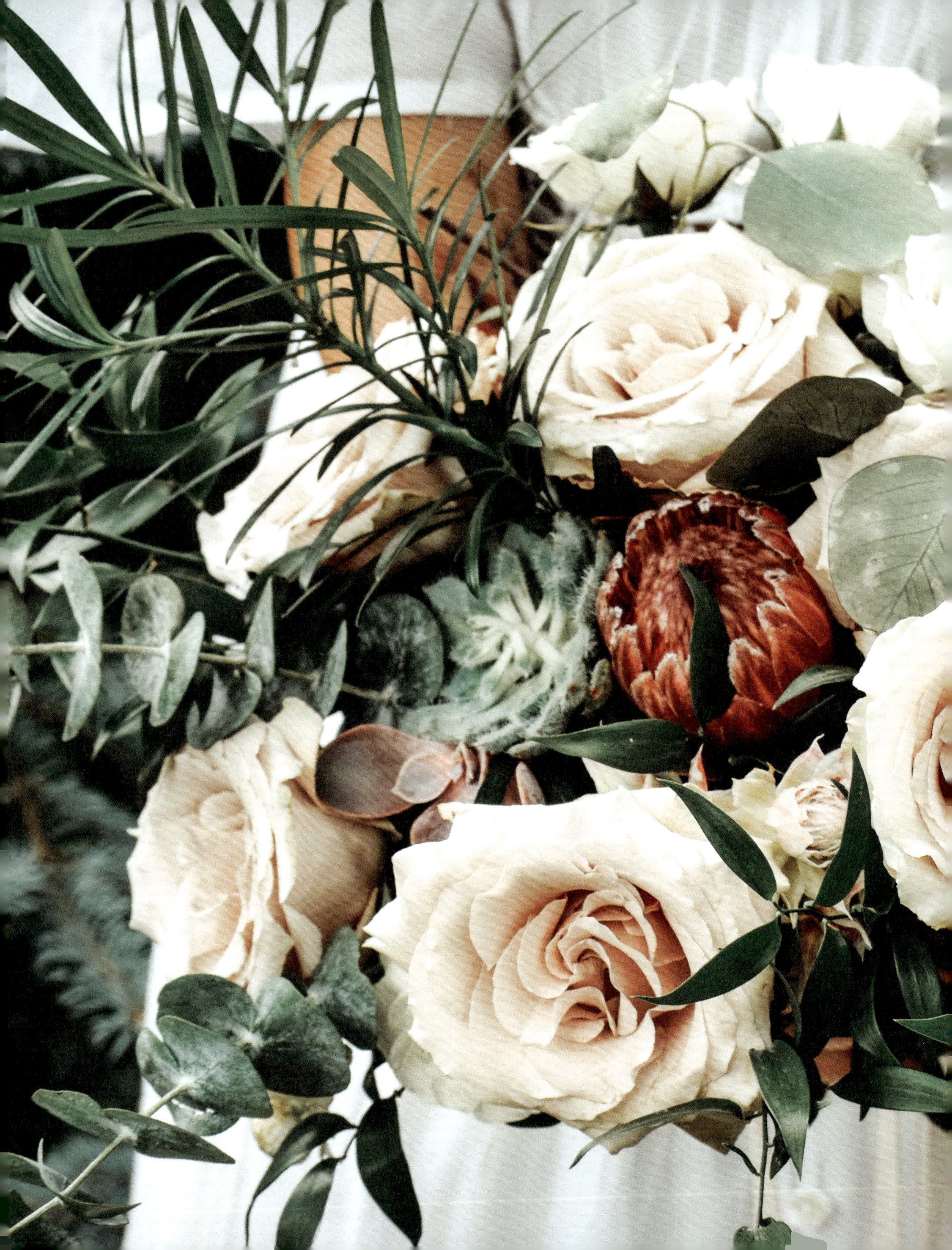

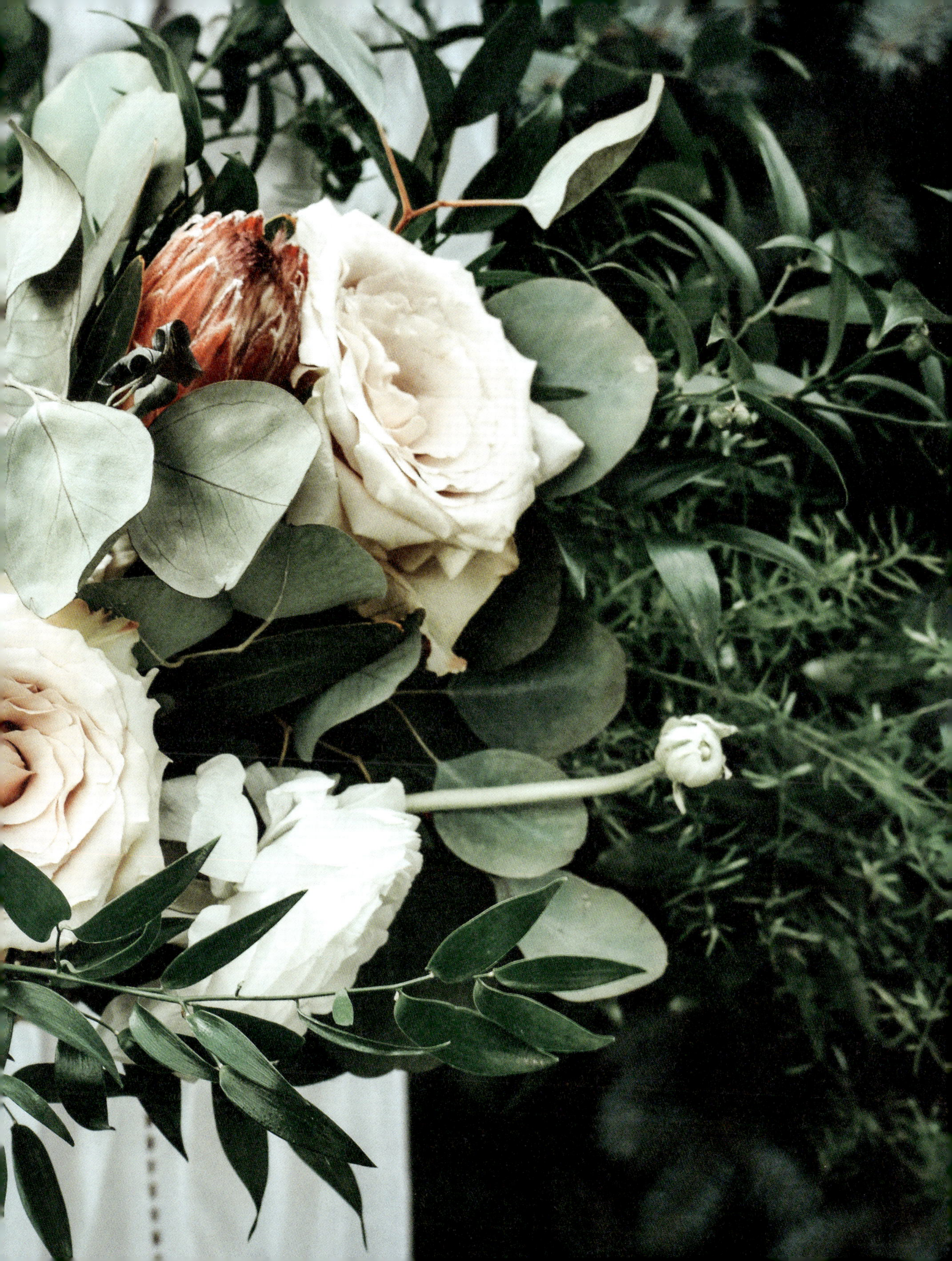

RED ROSES

LOVE AND PASSION

A red rose epitomizes romance and bespeaks passion, sensuality, and unconditional love. Valentine's Day or no, there is perhaps no more romantic unspoken declaration of love than a dozen long-stemmed red roses. The intense, satiny abundance revealed in the slowly unfolding petals of a red rose contain so much beauty and perfection—Who knows what might be in store after this blossoming "I love you"?

ROTE ROSEN

LEIDENSCHAFT UND LIEBE

Eine rote Rose ist der Inbegriff der Romantik und spricht von Leidenschaft, Sinnlichkeit und bedingungsloser Liebe. Ein Dutzend langstielige rote Rosen sind, nicht nur am Valentinstag, die wohl romantischste Liebeserklärung ohne Worte. So viel Schönheit und Perfektion stecken in der intensiven samtigen Üppigkeit, welche die sich langsam entfaltenden Blütenblätter einer roten Rose offenbaren – und wer weiß, was auf dieses aufblühende „Ich liebe dich" folgen wird …?

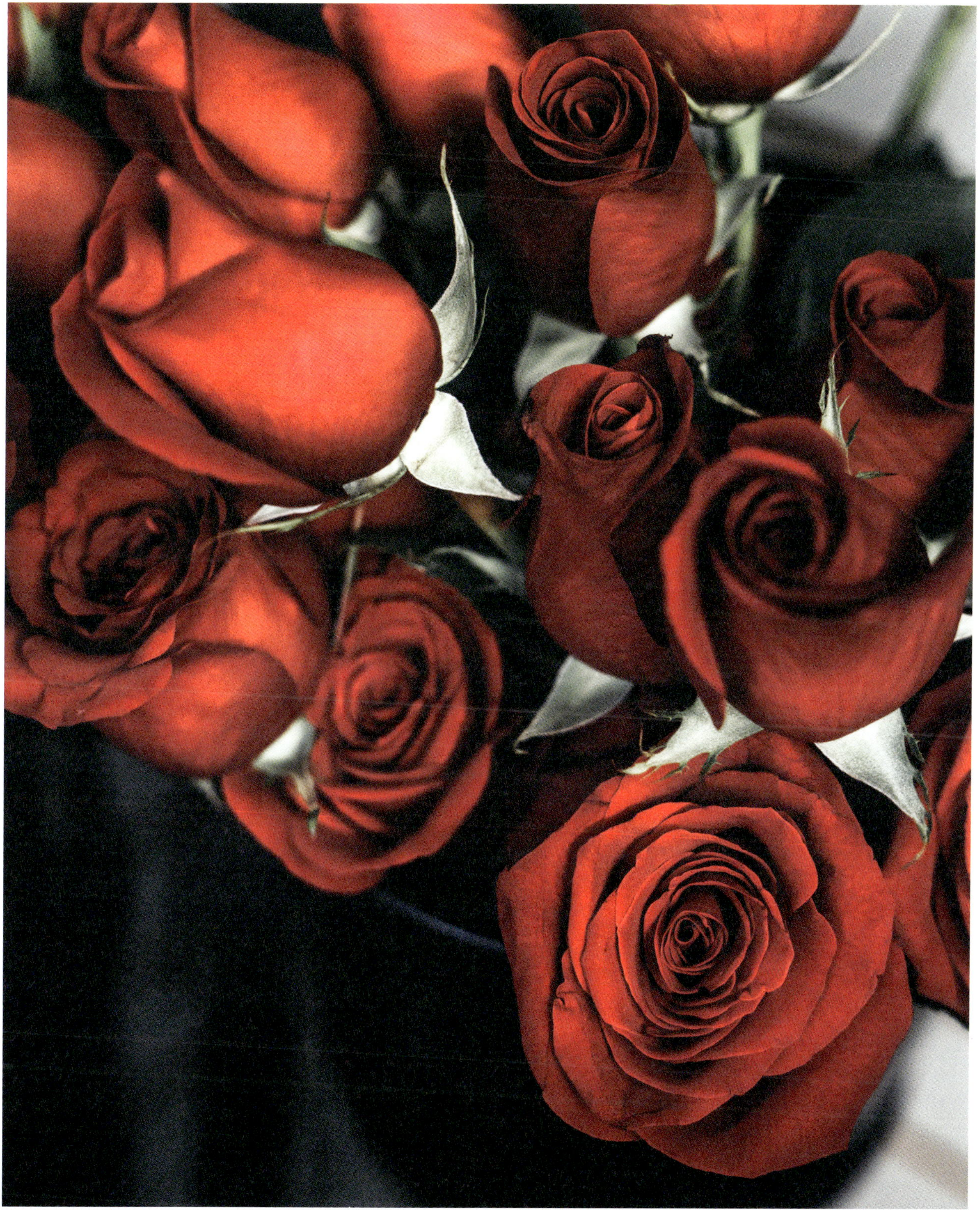

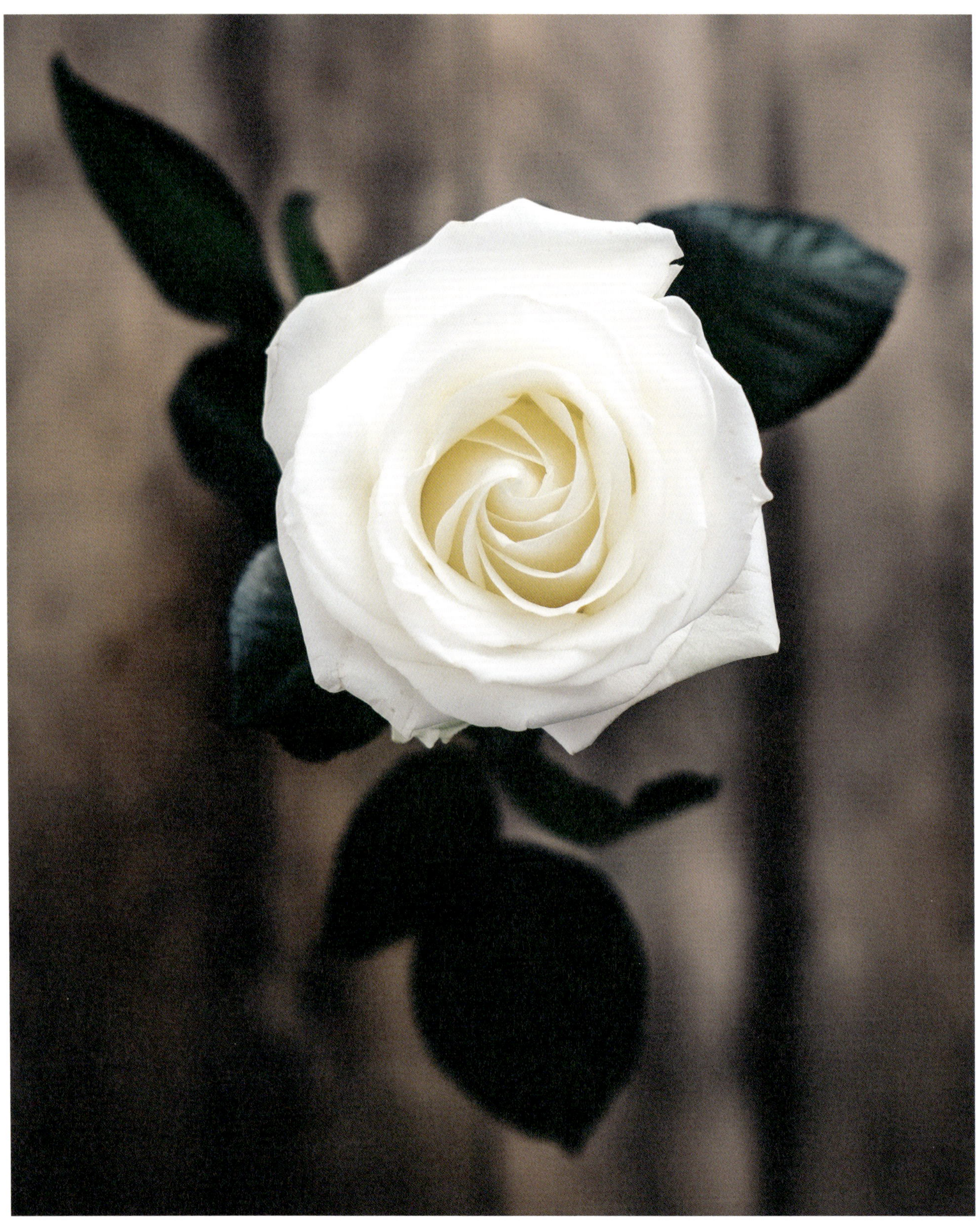

WHITE ROSES

INNOCENCE AND ELEGANCE

Noble white roses embody elegance and clarity. The color white is regarded as a symbol of innocence and purity. A delicate rose bouquet in virgin white goes well with almost any wedding dress. (No wonder the white rose is most popular for weddings.) But whether it's a new beginning or a farewell, white roses are the right choice for both a baptism celebration and as a final greeting for the deceased.

WEISSE ROSEN

UNSCHULD UND ELEGANZ

Edle weiße Rosenblüten sind die Verkörperung von Eleganz und Klarheit. Allgemein gilt die Farbe Weiß als Symbol für Unschuld und Reinheit. Ein zartes Rosenbouquet in jungfräulichem Weiß passt zu so gut wie jedem Hochzeitskleid – kein Wunder, dass die weiße Rose als beliebteste Hochzeitsblüte gilt. Doch ob Neubeginn oder Abschied, sowohl bei einer Tauffeier als auch als letzter Gruß für Verstorbene sind weiße Rosen die richtige Wahl.

HOT PINK

GRATITUDE AND ADMIRATION

Rose petals in an intense, bright pink are full of positive emotions. Whoever presents a bouquet of roses with lush pink blossoms expresses sincere admiration and appreciation in a flowery way. Also, who wants to convey feelings of happiness, joy, and gratitude? A beautiful, powerful bouquet of pink roses can express this quite elegantly, and without a hidden agenda.

PINKE ROSEN

BEWUNDERUNG UND DANKBARKEIT

Rosenblüten in intensivem leuchtendem Pink stecken voll positiver Emotionen: Wer einen Rosenstrauß mit üppigen pinkfarbenen Blüten überreicht, drückt damit blumig aufrichtige Bewunderung und ehrliche Anerkennung aus. Auch wer besondere Glücksgefühle, Freude und Dankbarkeit zeigen möchte, kann dies ganz elegant und ohne verborgene Hintergedanken mit einem wunderschönen Rosenbouquet in kraftvollem Pink zum Ausdruck bringen.

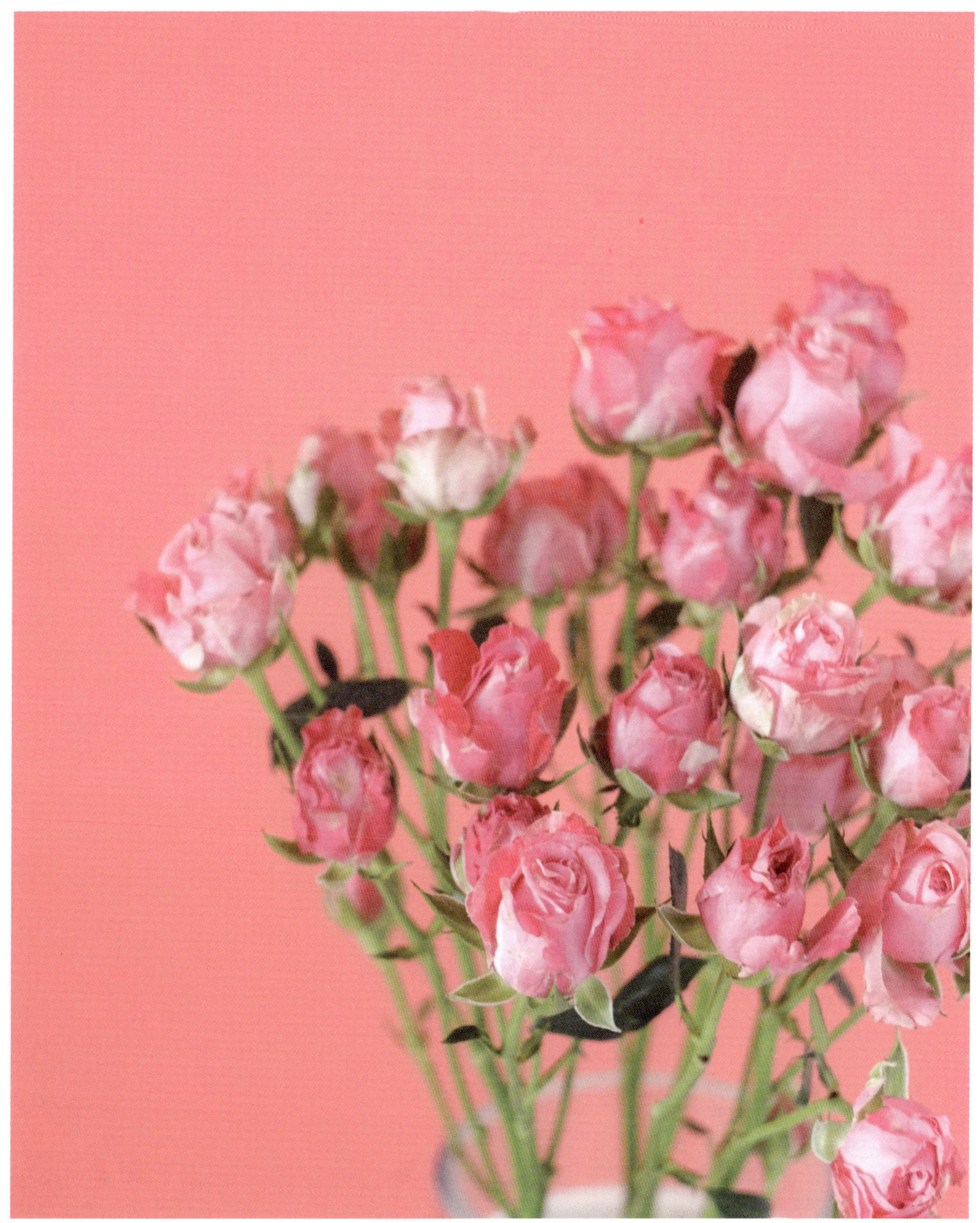

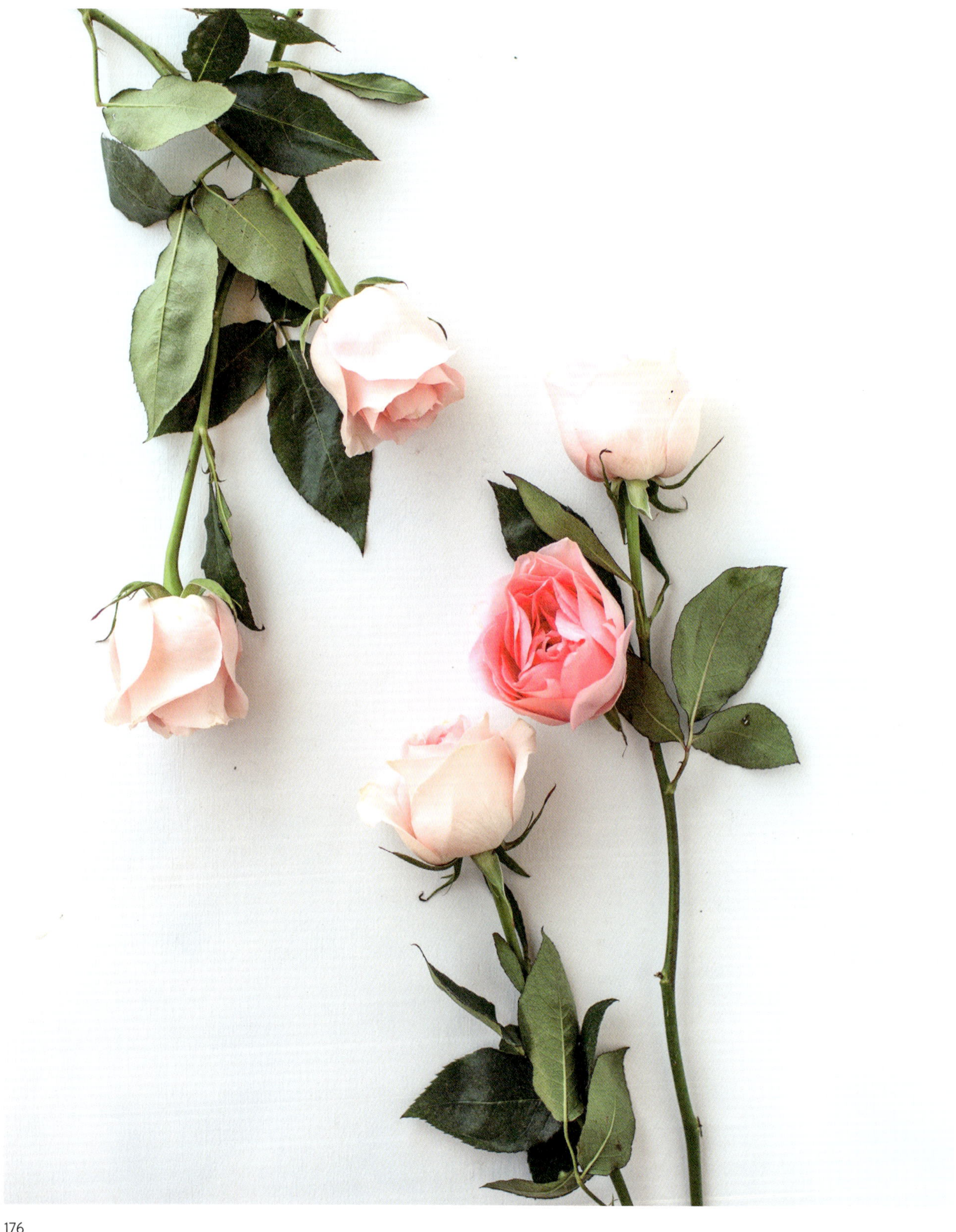

CARNATION PINK

GRACE AND BEAUTY

Young, fresh love, rapturous admiration, romantic adoration—a bouquet of delicate pink roses can be it all. The soft, unobtrusive hue of a flower in subtle pink is generally considered the epitome of grace and beauty. Pink forms a bridge between innocent white and passionate red: Where a young love is about to bloom, a bouquet of delicate pink rosebuds, kind of shyly, testifies to loving affection.

ROSAFARBENE ROSEN

ANMUT UND SCHÖNHEIT

Junge, frische Liebe, schwärmerische Bewunderung und romantische Anbetung – all das kann in einem Strauß von zart rosafarbenen Rosen stecken. Der sanfte, unaufdringliche Farbton einer Blüte in dezentem Rosa gilt allgemein als Inbegriff von Anmut und Schönheit. Rosa ist die Brücke von unschuldigem Weiß zu leidenschaftlichem Rot: Wo eine junge Liebe erst im Begriff ist zu erblühen, zeugt ein Bouquet von zartrosa Rosenknospen fast schüchtern von liebevoller Wertschätzung.

YELLOW ROSES

A SWIRL OF EMOTIONS

Sun-yellow roses radiate the wonderful warmth of a summer's day: Especially the delicate little buds of a moss-rose or centifolia, for example, convey positive emotions such as happiness and friendship. But beware, with a bold bouquet of long-stemmed yellow hybrid tea roses, an admirer can also flash his jealousy—or perhaps it is also an offer to reconcile after the passionately inflamed mind has come, again, to rest?

GELBE ROSEN

EIN WIRBEL DER EMOTIONEN

Sonnengelbe Rosen strahlen die wunderbare Wärme eines Sommertags aus: Besonders die zarten kleinen Knospen etwa einer Moosrose vermitteln positive Emotionen wie Glück und freundschaftliche Verbundenheit. Aber Achtung, mit einem kühnen Strauß langstieliger gelber Edelrosen kann ein Verehrer auch seine Eifersucht aufblitzen lassen – oder vielleicht ist es auch ein Angebot zur Versöhnung, nachdem sich das in Leidenschaft erhitzte Gemüt wieder beruhigt hat?

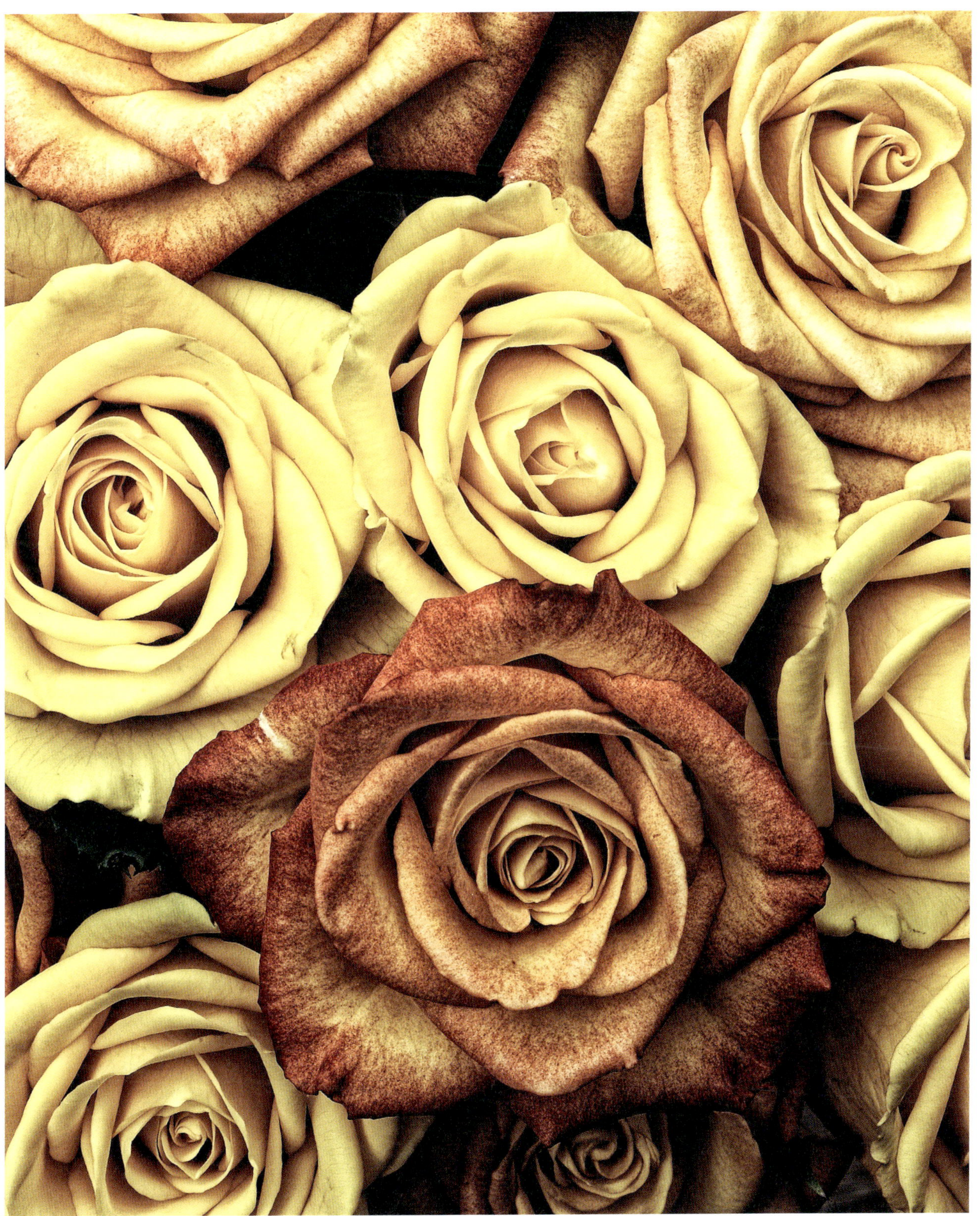

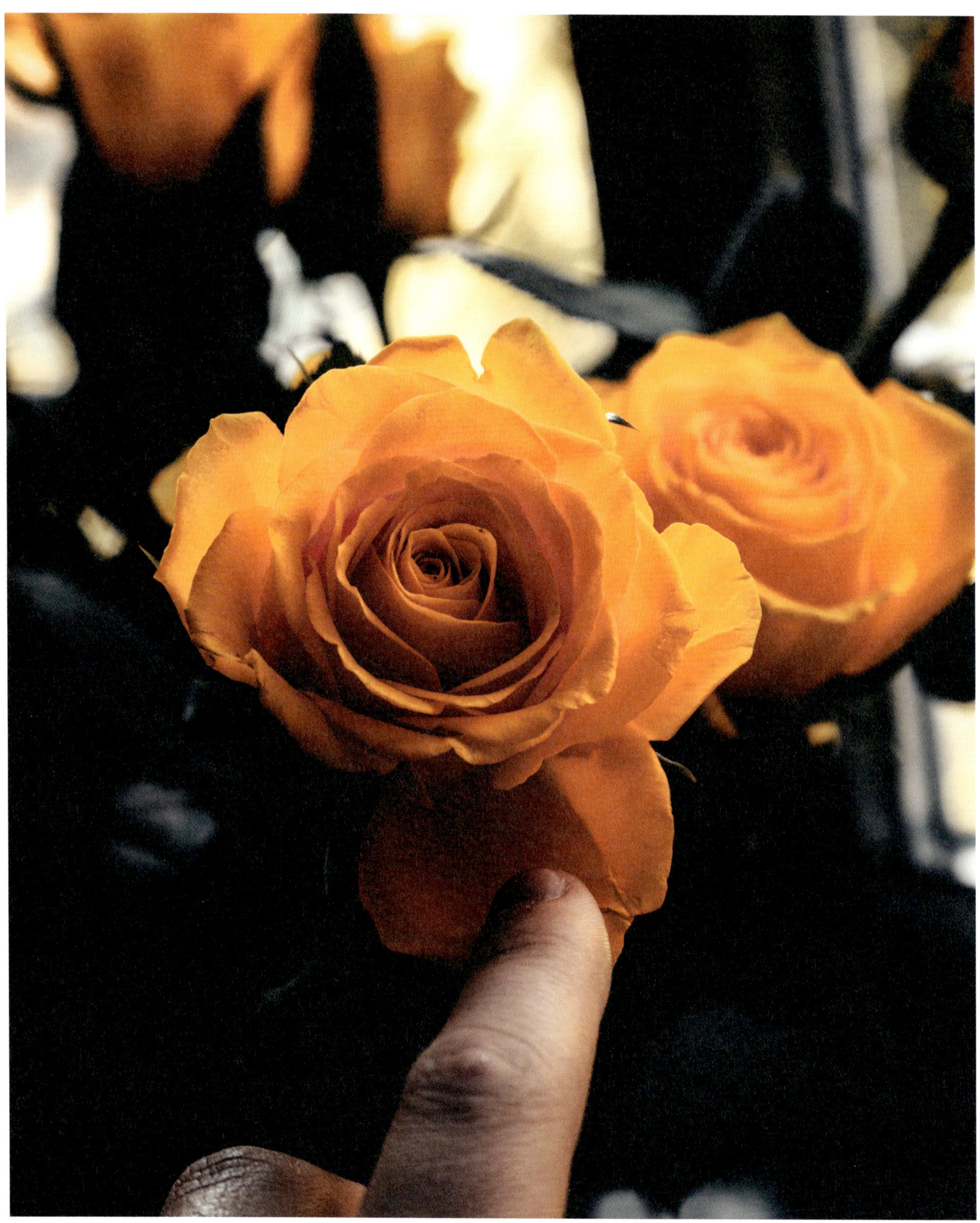

ORANGE ROSES

ENERGY AND FASCINATION

Bright orange rose petals are all enthusiasm and fascination. A bouquet of fiery orange roses transmits positive energy and joy. In a relationship, the bright flower speaks of happiness and contentment, of a wonderful time together, and may it also last. A beautiful, uplifting shade for get-well wishes, too: orange roses provide strength and vitality to any hospital room.

ORANGEFARBENE ROSEN

FASZINATION UND ENERGIE

Das leuchtende Orange von Rosenblüten steckt voller Enthusiasmus und Faszination. Wer einen Strauß feurig orangefarbener Rosen überreicht, übermittelt damit positive Energie und Freude. In einer Beziehung sprechen die leuchtenden Blüten von Glück und Zufriedenheit – so wunderbar darf die gemeinsame Zeit noch lange weitergehen! Ein schöner, aufmunternder Farbton auch für Genesungswünsche: Rosen in Orange bringen Kraft und Lebensfreude in jedes Krankenzimmer.

PEACH

GRATITUDE AND SYMPATHY

Giving a soft, warm peach shade of roses signals gratitude and appreciation. Especially when a couple is still new, the parties unsure of whither the journey, shades of peach in a bouquet of roses set a positive sign of affection— and is this perhaps the prelude to more? In a relationship, on the other hand, they let the recipient know that the partner feels secure and comfortable together.

PFIRSICHFARBENE ROSEN

DANKBARKEIT UND SYMPATHIE

Rosen in einem sanften, warmen Pfirsichton zu schenken signalisiert Dankbarkeit und Wertschätzung. Gerade wenn sich ein Paar erst kennengelernt hat und man noch nicht ganz weiß, wohin die Reise gehen wird, setzt ein Rosenstrauß in Pfirsichtönen ein positives Zeichen der Sympathie – und ist vielleicht der Auftakt zu mehr? Und wer in einer Beziehung ein Bouquet mit pfirsichfarbenen Rosen erhält, kann daran erkennen, dass der Partner sich in der Zweisamkeit wunderbar geborgen fühlt.

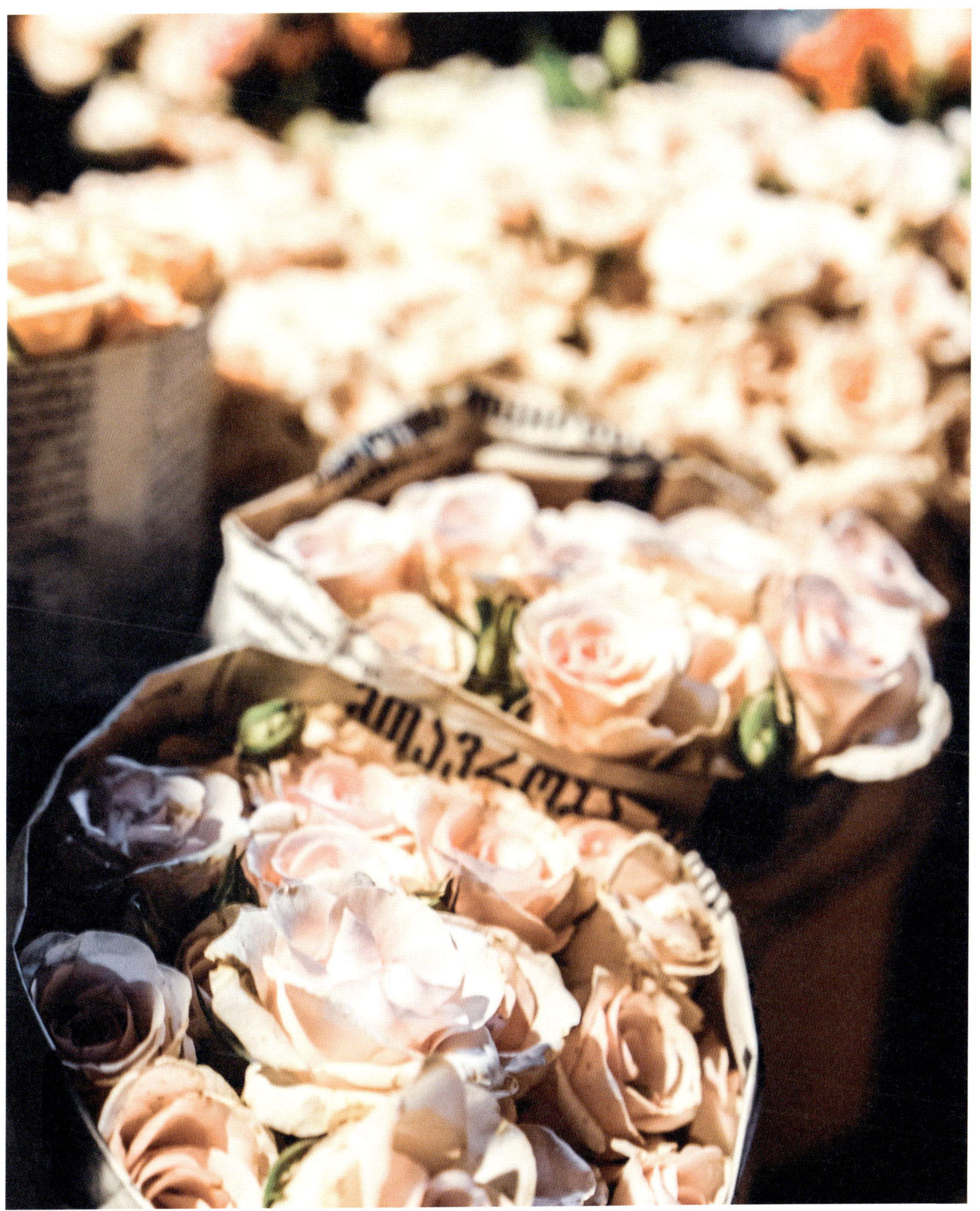

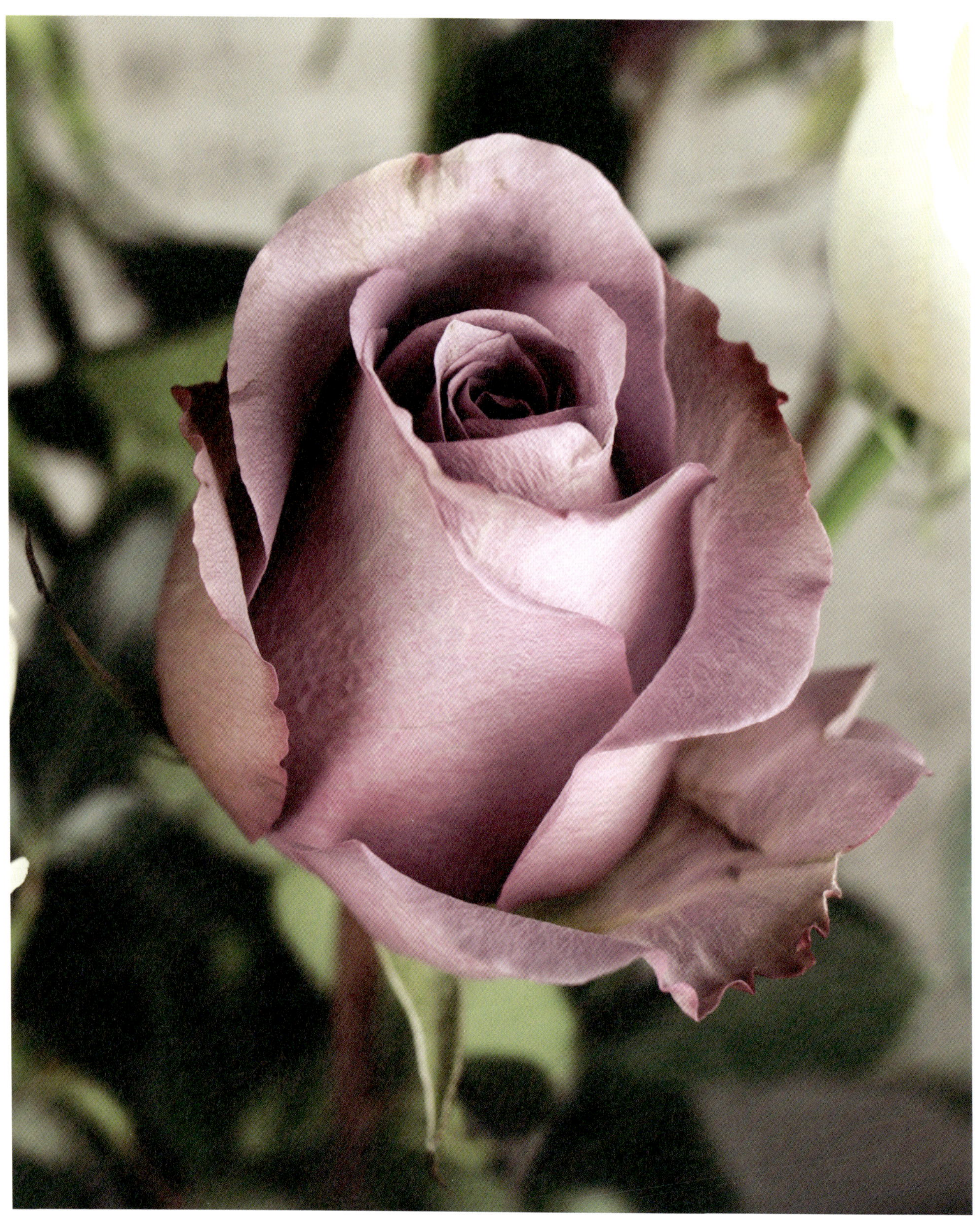

PURPLE ROSES

MAGIC AND DISTINCTION

Purple-flowered roses are rare in nature—and that's exactly what makes them magical. The soft touch of lavender and lilac enchants us and touches us peculiarly. A purple bouquet of roses testifies to love at first sight: "I have found the only person in the whole world to whom my heart shall belong—no-one other than *you* on my mind!"

VIOLETTE ROSEN

EINZIGARTIGKEIT UND ZAUBER

Rosen, die in Violett erblühen, sind in der Natur eine Seltenheit – und genau das macht ihre Magie aus: Der sanfte Farbhauch von Lavendel und Flieder bezaubert und berührt uns auf eigentümliche Weise. Ein violetter Rosenstrauß zeugt von Liebe auf den ersten Blick: „Ich habe den einzigen Menschen auf der ganzen Welt gefunden, dem mein Herz gehören soll – an nichts anderes kann ich mehr denken als nur an dich!"

COLORFUL

AN INSPIRING ZEST FOR LIFE

Colorful and carefree is how roses bloom in nature. Even in a bouquet, a bright ensemble of roses gives joy and variety. A colorful bouquet combining lush flowers of every nuance radiates *joie de vivre* and baits the recipient to smile. And for those who want to convey some joy to a flower-lover without unintentionally sending a secret message, multi-colored rose combinations are always a safe option.

BUNTE ROSEN

INSPIRIERENDE LEBENSFREUDE

Bunt gemischt und unbekümmert – so blühen Rosen in der Natur. Und auch in einem Bouquet schenkt ein farbenfrohes Ensemble Freude und Abwechslung. Ein bunter Rosenstrauß, der üppige Blüten in allen Nuancen in sich vereint, strahlt pure Lebensfreude aus und zaubert dem Beschenkten immer ein Lächeln ins Gesicht. Und wer Blumenfreunden einfach eine Freude machen möchte, ohne ungewollt eine geheime Botschaft zu übermitteln, für den sind mehrfarbig kombinierte Rosen immer eine sichere Wahl.

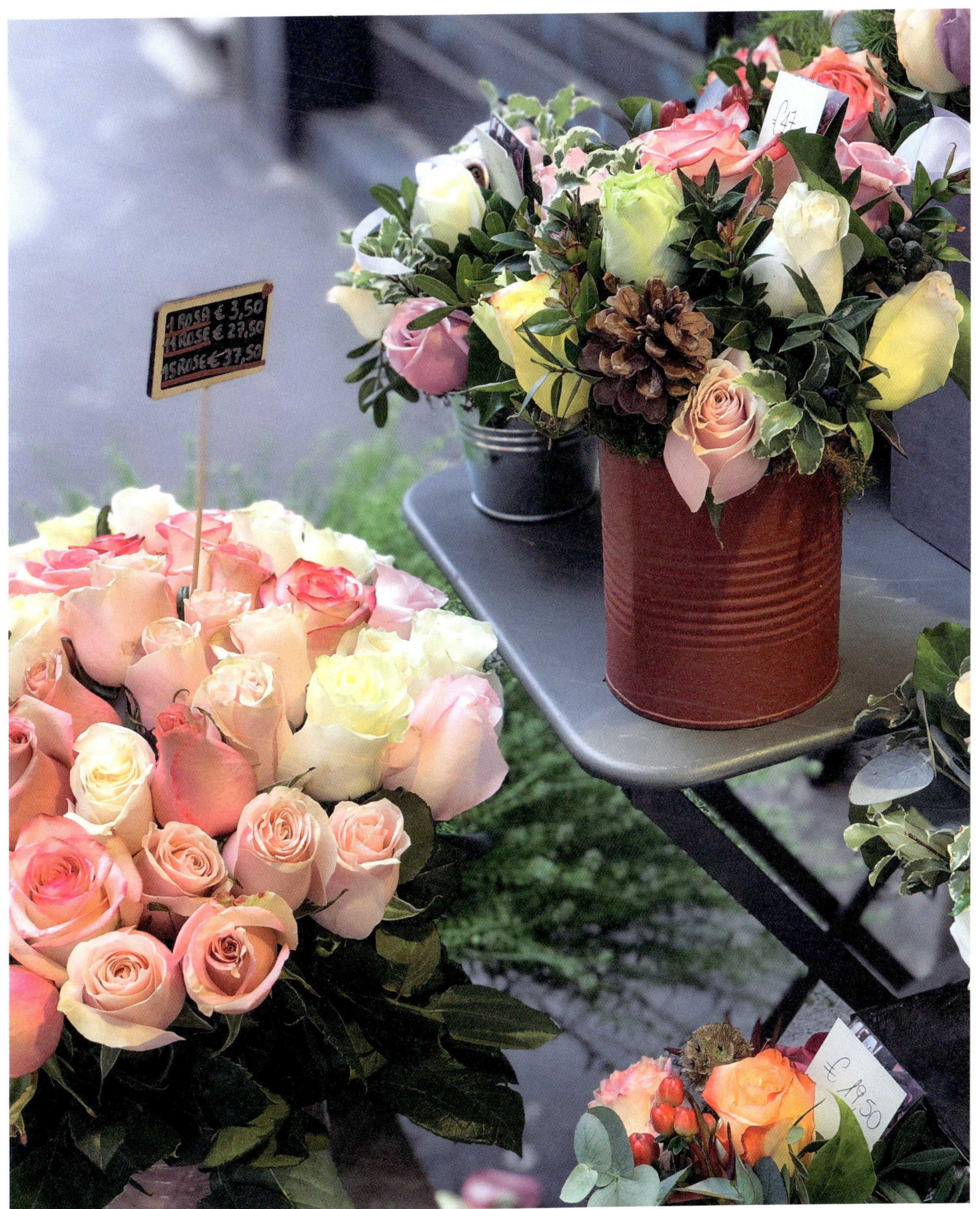

THE RO
IN ART

DIE ROSE
IN DER

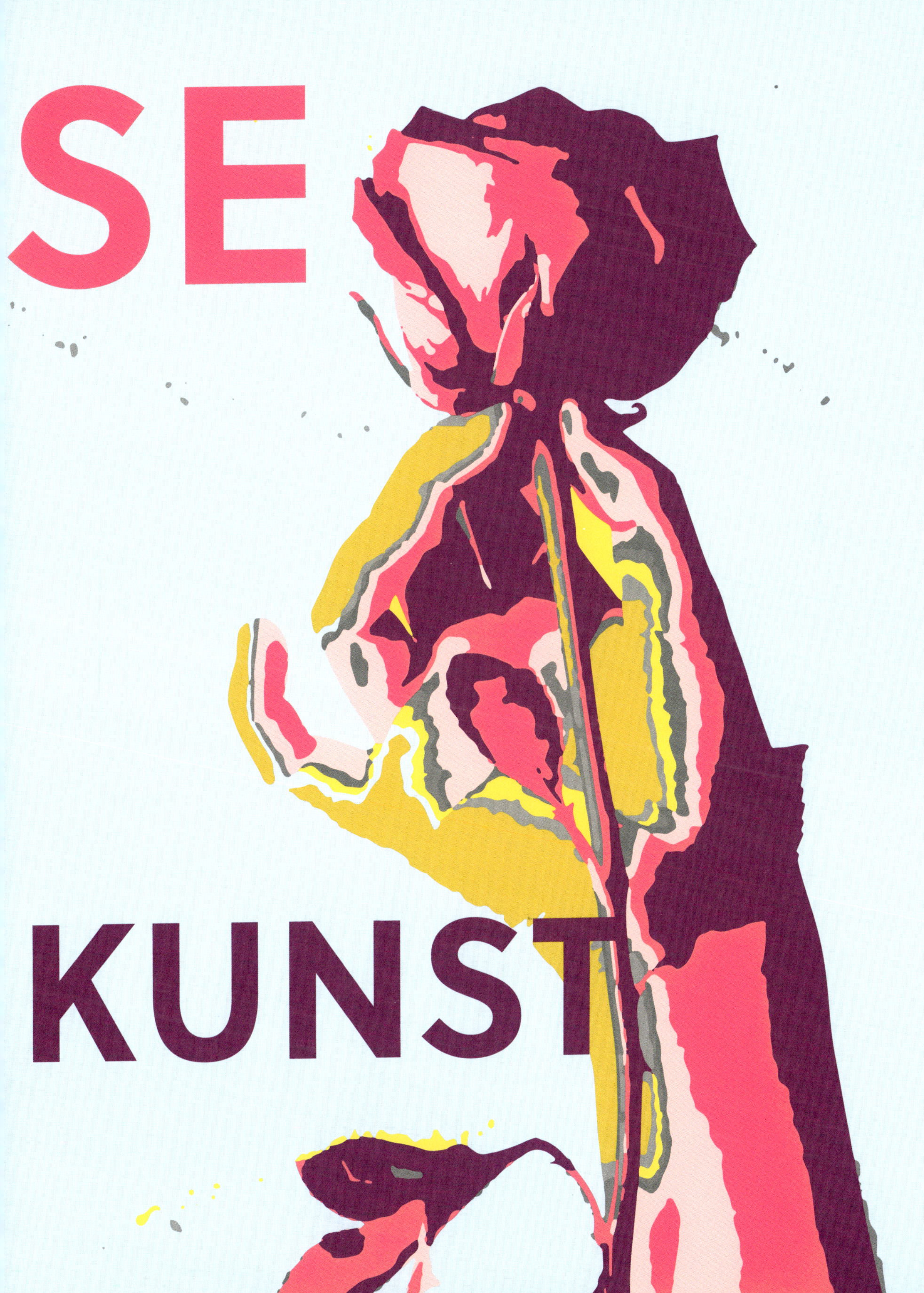

SE
KUNST

Jan Davidszoon de Heem

MEMENTO MORI

With *Memento Mori*, the seventeenth-century Dutch painter Jan Davidszoon de Heem entered the tradition of vanitas still lifes. His magnificent compositions, with flowers and fruits, were considered to be allegorical representations of the motif of transience. His wealthy patrons did not always accept the covert symbolism—but what clearer reminder of the finiteness of earthly life could there be than a skull and a sheet of paper bearing the words "Memento mori"?

MEMENTO MORI

Mit *Memento Mori* reihte sich der niederländische Maler Jan Davidszoon de Heem im 17. Jahrhundert in die Tradition der Vanitas-Stillleben ein. Die prunkvollen Kompositionen mit Blumen und Früchten galten als allegorische Darstellungen zum Motiv der Vergänglichkeit. Die wohlhabenden Auftraggeber waren nicht immer aufgeschlossen für die verborgene Symbolik – doch deutlicher als mit der Abbildung eines Totenschädels und den Worten „Memento mori" auf einem Blatt Papier konnte man die Endlichkeit des irdischen Lebens kaum anmahnen.

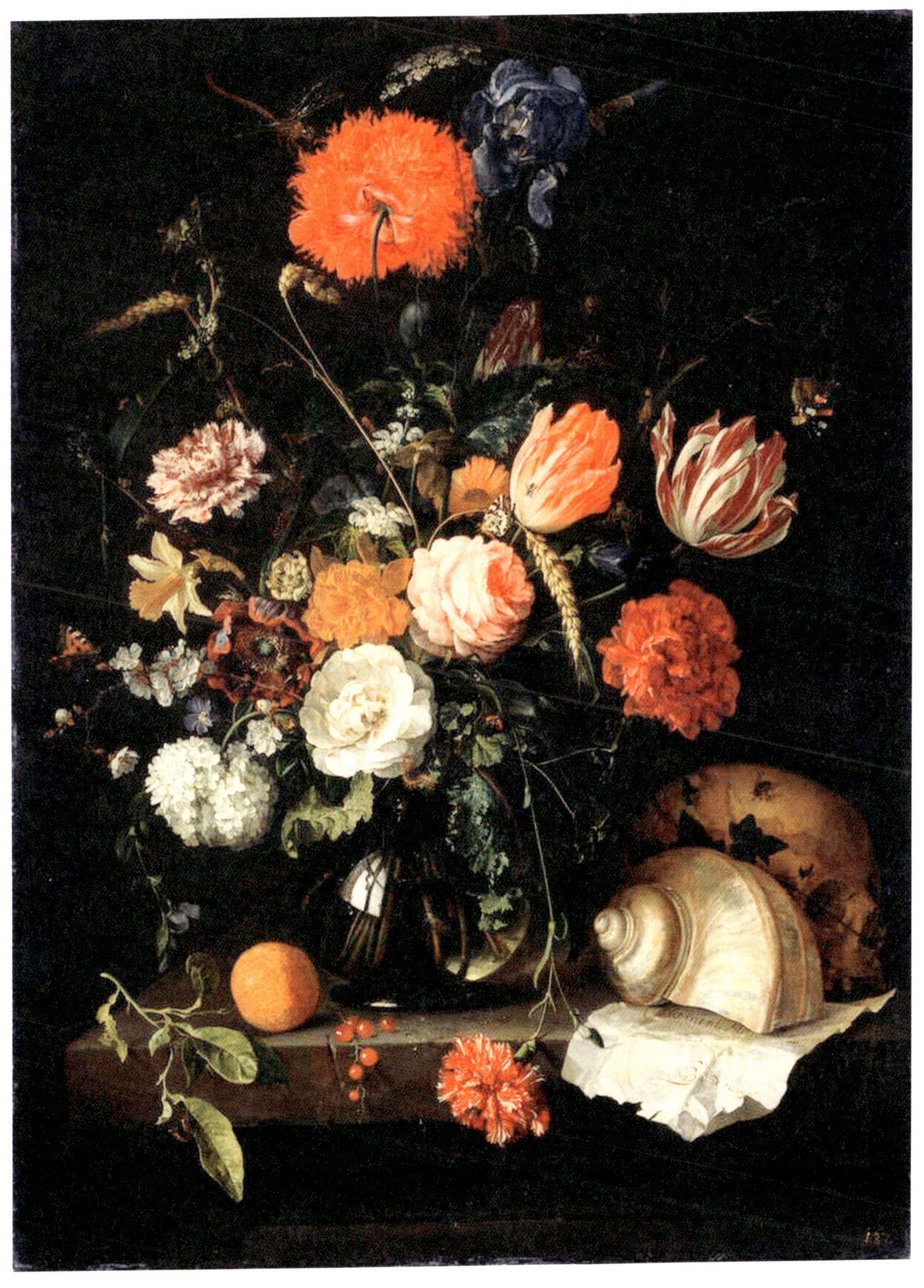

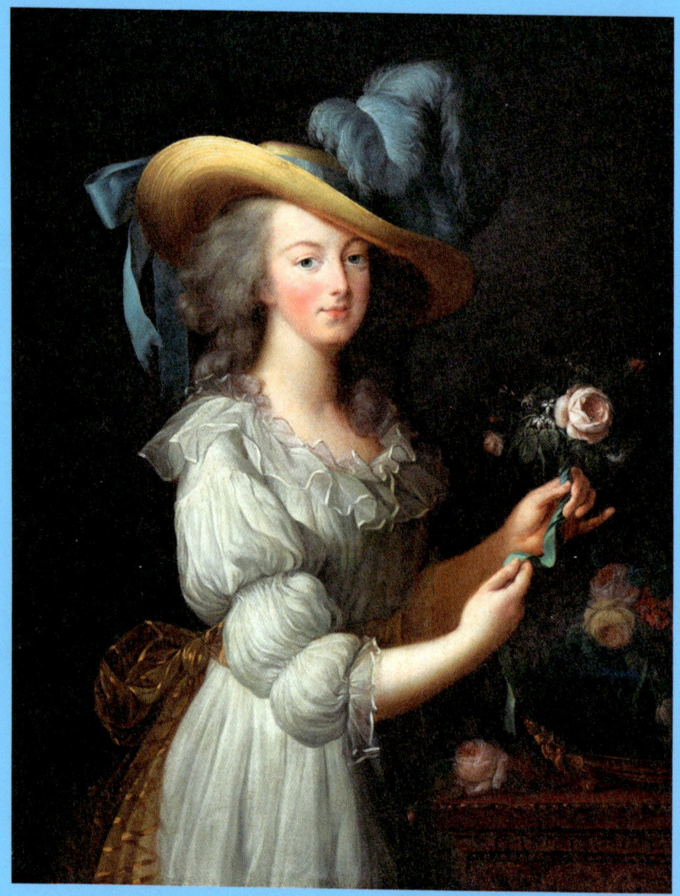

MARIE-ANTOINETTE WITH A ROSE

Élisabeth Vigée Le Brun portrayed Queen Marie-Antoinette with her favorite rose at the Paris Salon in 1783, six years before the French Revolution—and immediately created a minor scandal. How can you show the empress robed in what looks like an undergarment?

Madame Le Brun, whose career and income were based on doing portraiture work in aristocratic circles, reacted promptly, presenting a follow-up portrait of the queen while the salon was still in progress—this time in a stylish formal gown.

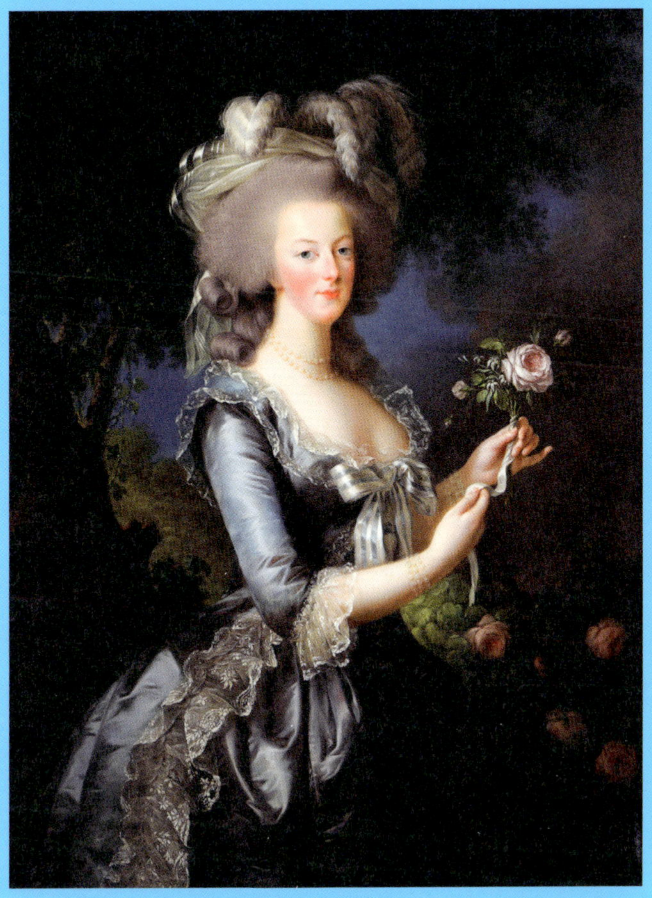

MARIE-ANTOINETTE MIT EINER ROSE

Im Jahr 1783, sechs Jahre vor dem Beginn der Französischen Revolution, schuf Élisabeth Vigée Le Brun anlässlich des Pariser Salons ein Gemälde von Königin Marie-Antoinette mit ihrer Lieblingsrose (linke Seite) – und sorgte damit gleich für einen kleinen Skandal: Wie kann man nur die Königin in einem Gewand zeigen, das wie ein Unterkleid aussieht! Madame Le Brun, deren Karriere und Einkommen auf Porträts in Adelskreisen gründete, reagierte prompt und präsentierte noch während des Salons das Nachfolgewerk – diesmal mit formaler, schicklicher Robe (diese Seite).

Vincent van Gogh

ROSES

One of Vincent van Gogh's greatest still lifes is from 1890, shortly before the artist left the sanatorium at Saint-Rémy. For Van Gogh, perceiving nature in full bloom was essential to maintaining his health—or to regaining it—at the sanitarium. The leaf green of the bouquet of roses in the vase harmonizes beautifully with the luminescent pale- and tea-green background. The fact that van Gogh painted the flowers in an almost ethereal white, although they were pink, may be an indication of how exhausted he was from his illness.

ROSEN

Eines der größten Stillleben Vincent van Goghs entstand 1890, kurz bevor der Künstler das Sanatorium in Saint-Rémy verließ. Die Natur in voller Blüte wahrzunehmen hielt van Gogh für unerlässlich, um seine Gesundheit zu erhalten – oder im Sanatorium wiederzuerlangen. Mit dem hellgrün leuchtenden Hintergrund harmoniert das Blattgrün des Rosenstraußes in der Vase wunderbar. Dass van Gogh die eigentlich rosafarbenen Blüten aber in fast ätherischem Weiß malte, mag ein Hinweis sein, wie sehr seine Krankheit ihn ausgelaugt hatte.

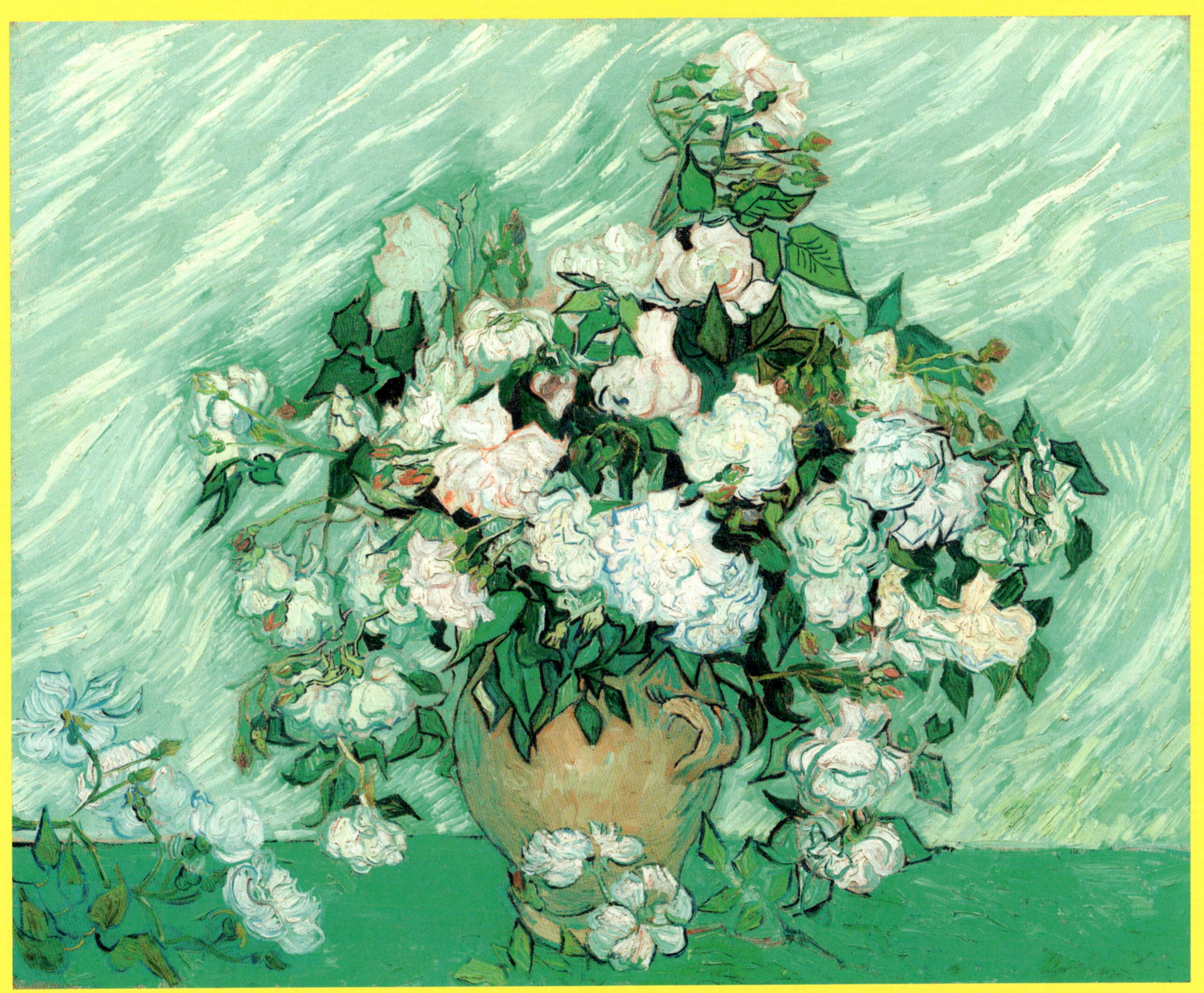

John William Waterhouse

THE SOUL OF
THE ROSE

John William Waterhouse was committed to the Pre-Raphaelite tradition of combining the enchantment of nature with pure feminine beauty. One line, from Alfred Lord Tennyson's poem *Maud*, particularly inspired Waterhouse: "And the soul of the rose passed into my blood." In the romanticizing vignette from the rose garden, you can almost feel the soul of the rose arising with its beguiling fragrance …

DIE SEELE DER ROSE

John William Waterhouse verschrieb sich der Tradition der Präraffaeliten, den Zauber der Natur und die reine weibliche Schönheit miteinander in Verbindung zu bringen. Eine Zeile aus Alfred Lord Tennysons Gedicht *Maud* inspirierte Waterhouse besonders: „Und die Seele der Rose ging auf in mein Blut." Und tatsächlich, in der romantisierenden Momentaufnahme aus dem Rosengarten kann man fast spüren, wie die Seele der Rose in ihrem betörenden Duft aufgeht …

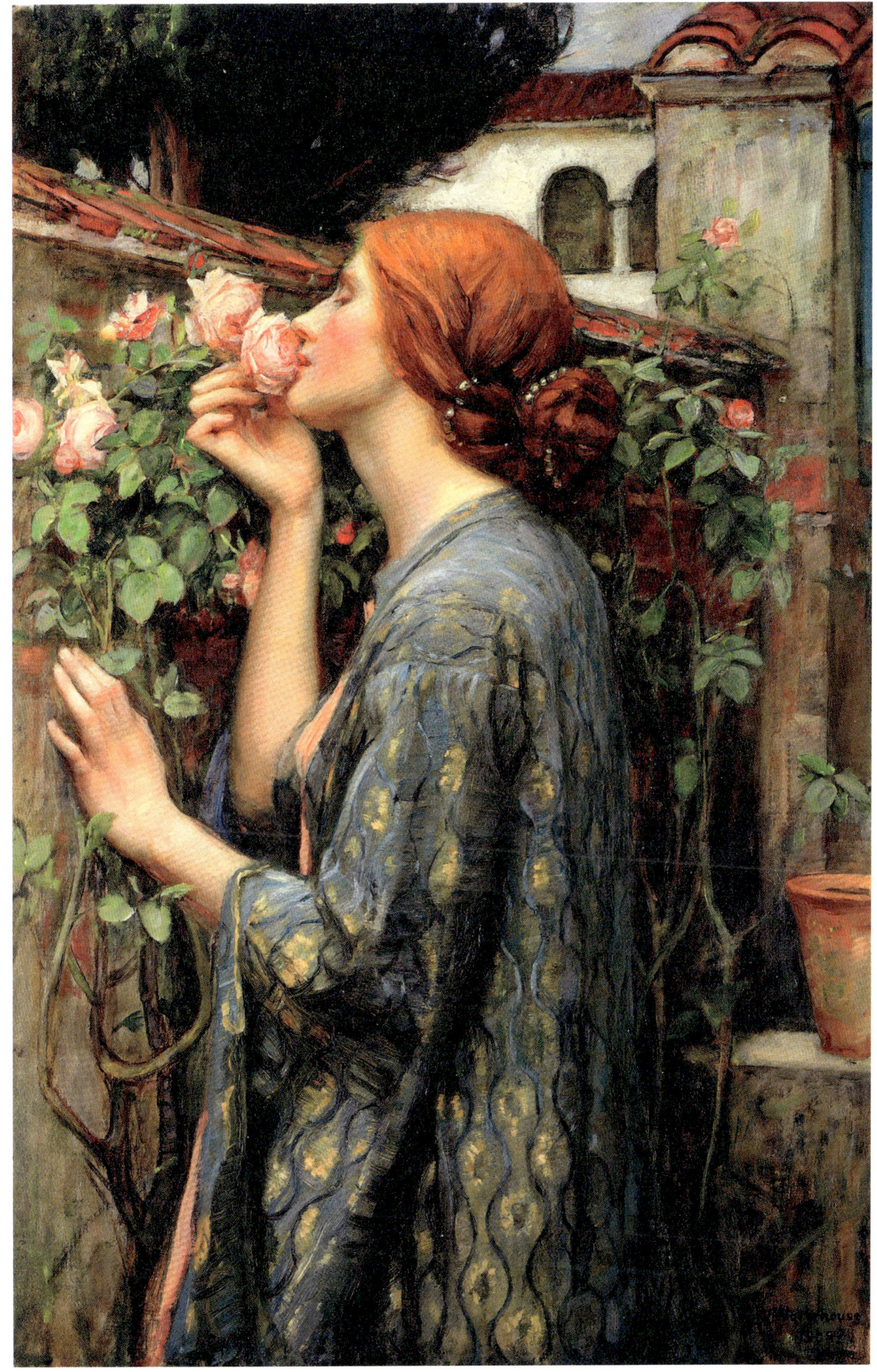

Georgia O'Keeffe

ABSTRACTION WHITE ROSE

Georgia O'Keeffe gets so close to her floral subjects (such as the white rose) that they lose their representational character and disintegrate into an abstraction of forms, colors, light, and shadow. As we plunge into its center, the blossom becomes unrecognizable; the viewer is forced to stop and query his own perception. And by the way: Throughout her life, O'Keeffe dismissed critics who sought to interpret her 1927 painting *Abstraction White Rose*, along with her other flower paintings, as expressions of her sexuality.

ABSTRAKTION WEISSE ROSE

So nah zog Georgia O'Keeffe ihre Blumenmotive oft an sich heran, dass sie ihre Gegenständlichkeit verloren und ganz in der Abstraktion von Farben, Formen, Licht und Schatten aufgingen. Indem sie quasi mitten in die Blüte eintaucht, wird diese nicht mehr als solche erkennbar: Sie zwingt den Betrachter, innezuhalten und die eigene Wahrnehmung zu hinterfragen. Den Kritikern, die *Abstraktion Weiße Rose* 1927 ebenso wie O'Keeffes andere Blütenbilder als Ausdruck ihrer Sexualität interpretieren wollten, erteilte die Malerin übrigens zeitlebens eine Abfuhr.

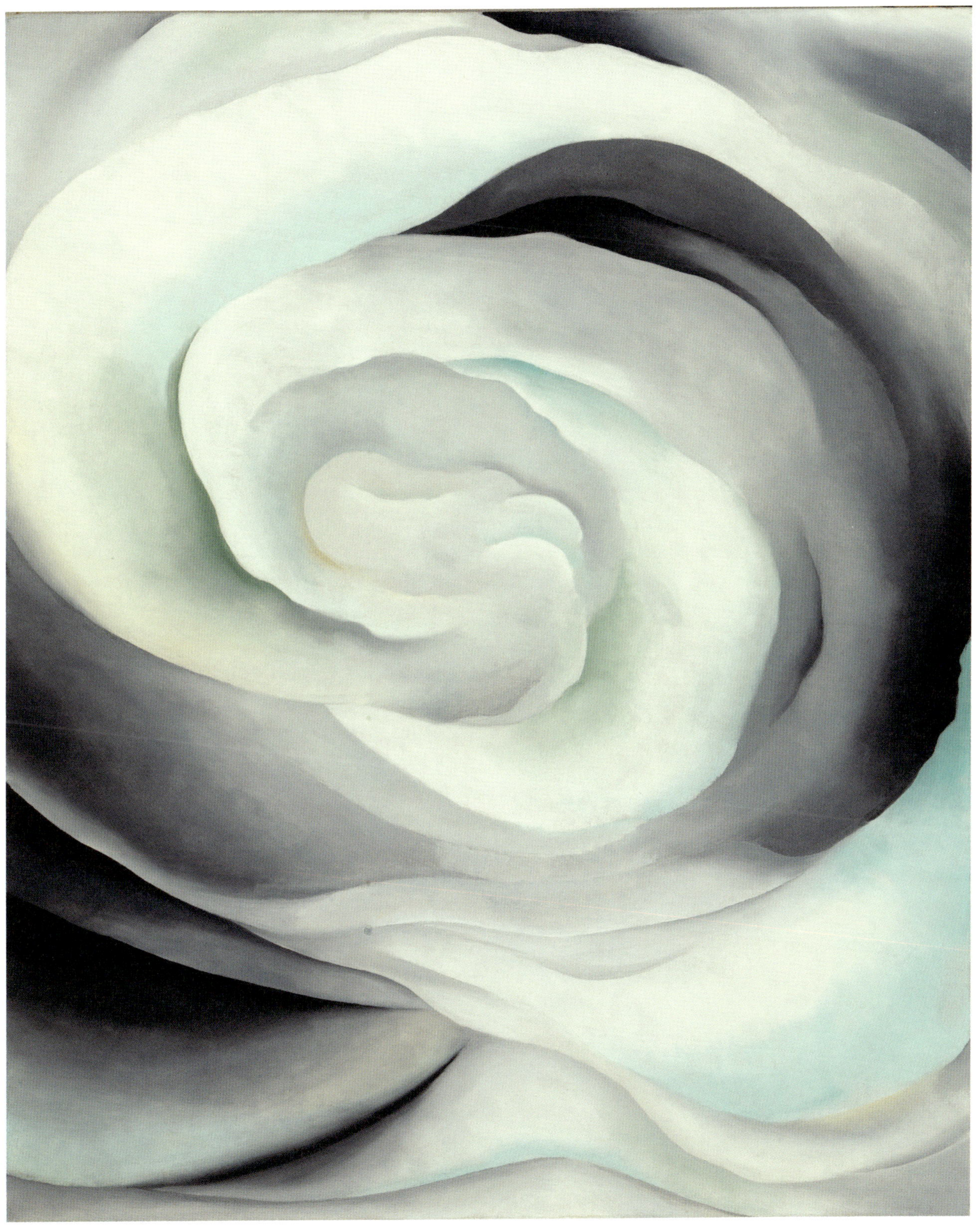

René Magritte

THE BACK OF A MAN WITH A ROSE

To paint in as many layers as there are registers in a language: this was the aim of the Belgian surrealist René Magritte. His enigmatic images turned everyday objects into a paradox of perception, into an empty shell of that which is represented. And so the rose, too, found its way into Magritte's series of motifs of the man in the suit and bowler hat. One could almost read into it a defiant contradiction of that famous line from a poem by Gertrude Stein: "A rose is a rose is a rose is a rose"…

DER RÜCKEN EINES MANNES MIT EINER ROSE

Malerei so vielschichtig zu machen wie Sprache – darauf zielte der belgische Surrealist René Magritte ab. Seine rätselhaften Bilder machten Alltagsobjekte zum Paradoxon der Wahrnehmung, das Dargestellte zur leeren Hülle. Und so hielt auch die Rose Einzug in Magrittes Motivreihe des Mannes mit Anzug und Melone: Fast könnte man darin einen trotzigen Widerspruch zu Gertrude Steins berühmter Gedichtzeile „Eine Rose ist eine Rose ist eine Rose ist eine Rose" lesen …

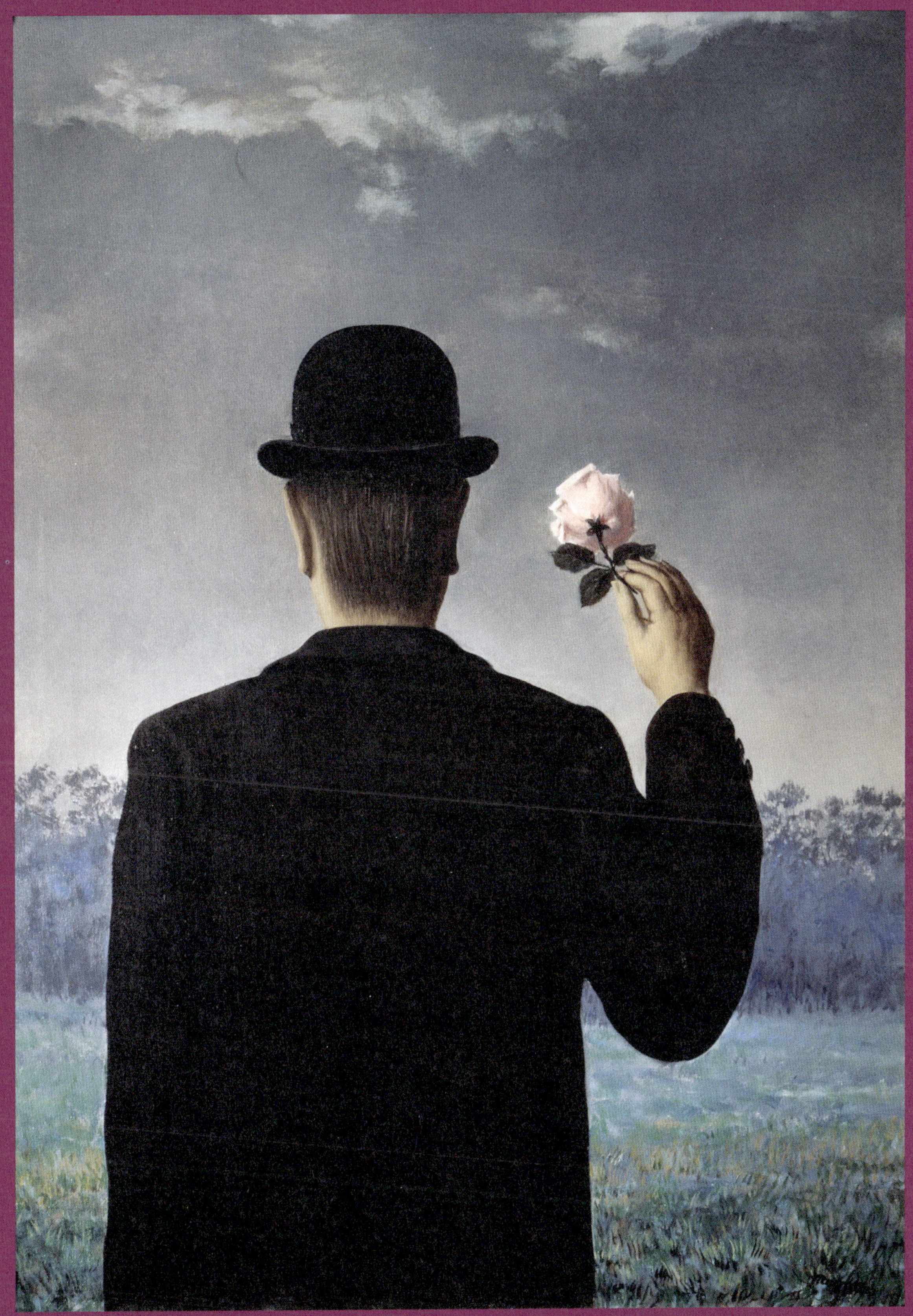

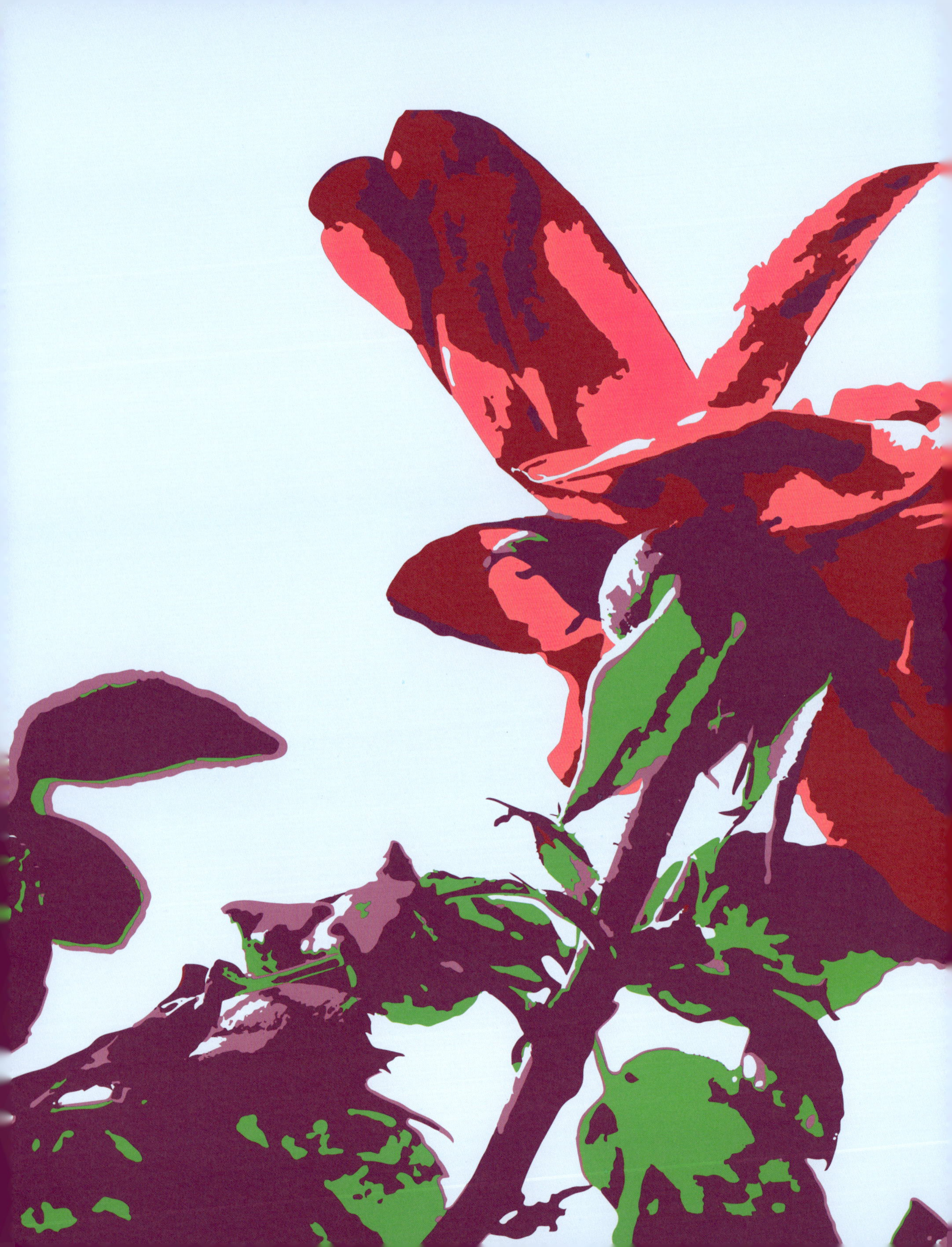

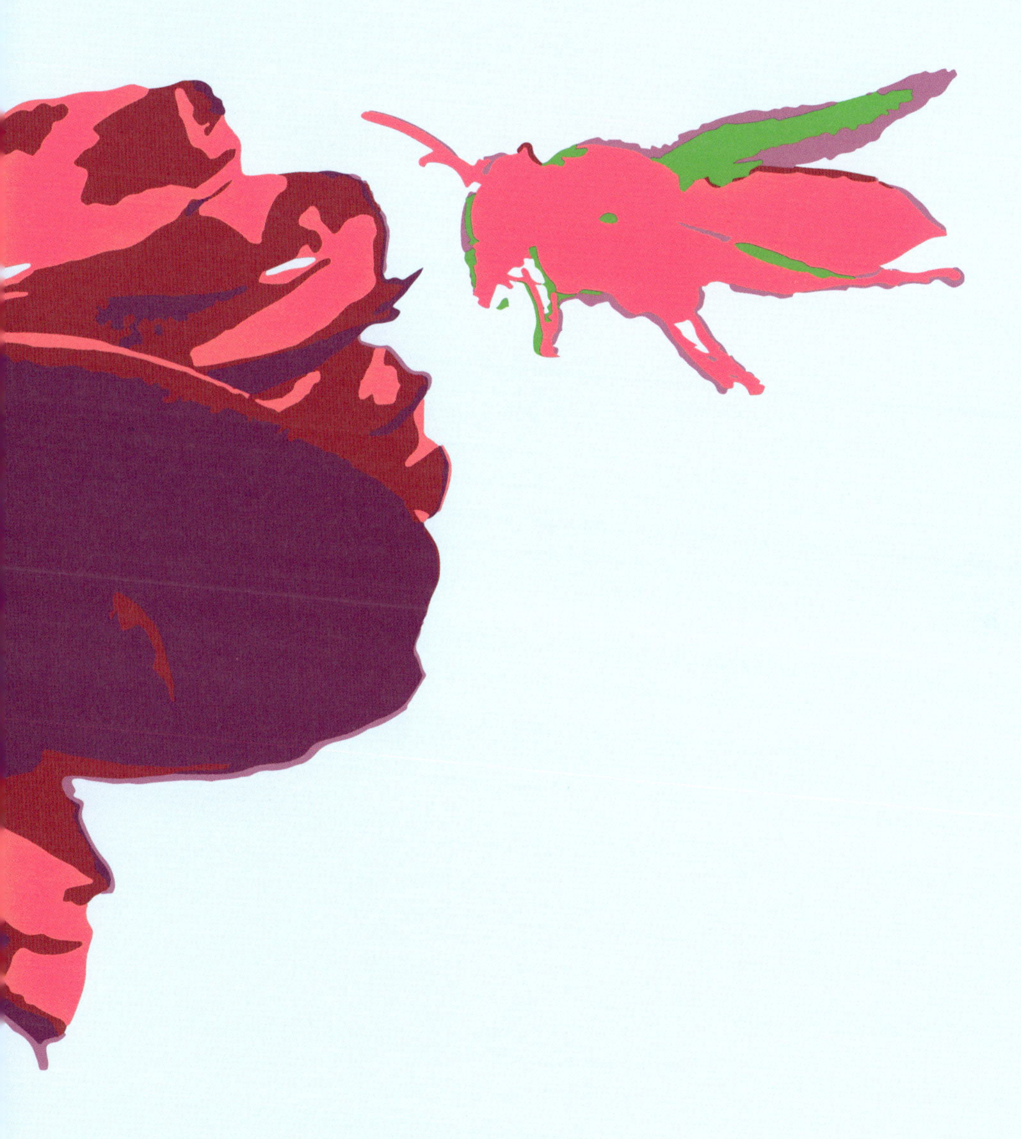

p. 104/105 Artefact/Alamy/Alamy Stock Photos/mauritius images, p. 107 top Vera Petruk/shutterstock.com, p. 107 bottom Gillian Singleton/Alamy/Alamy Stock Photos/mauritius images p. 108/109 Ricardo Gomez Angel/Unsplash, p. 111 Bridgeman Images, p. 113 top Markku murto/art/Alamy/Alamy Stock Photos/mauritius images, p. 113 bottom Rixie/stock.adobe.com, p. 114 National Museums Liverpool/Bridgeman Images, p. 116 The Picture Art Collection/Alamy/Alamy Stock Photos/mauritius images, p. 118 Tim Graham/gettyimages, p. 121 SuperStock/3LH-Fine Art/mauritius images, S.122 Doug Peters/Alamy/Alamy Stock Photos/mauritius images, S.125 Bourne Gallery, Reigate, Surrey/Bridgeman Images, p. 126/127 Lukyanova Elena/shutterstock.com, p. 128 Roka/shutterstock.com, S.130/131 Meg MacDonald/Unsplash, p. 132 Look and Learn/Bridgeman Images, p. 134 SuperStock/mauritius images, S.138 Mary Evans/The Pictures Now Image Collection/Interfoto, p. 139–146 Memento/mauritius images, p. 147 Rawpixel.com/stock.adobe.com, p. 149 picture alliance/Presse-Bild-Poss | Oscar Poss, p. 151 umarali804/shutterstock.com, p. 154/155 MasterChefNobu/shutterstock.com, p. 157 Nick Andros/shutterstock.com, p. 158/159 Catherine Southhouse/shutterstock.com, p. 160/161 Illustration: Eva Stadler, p. 162/163 WR.LILI/stock.adobe.com, p. 164/165 Frantisek Duris/Unsplash, p. 167 M.Khebra/shutterstock.com, p. 168/169 Jacalyn Beales/Unsplash, p. 171 Sidney Pearce/Unsplash, p. 172 Annie Spratt/Unsplash, p. 175 Aleksandra Sapozhnikova/Unsplash, p. 176 Rebecca nUg/Unsplash, p. 179 lobostudio hamburg/Unsplash, p. 180 Sebastiaan Stam/Unsplash, p. 183 Erinada Valpurgieva/Unsplash, p. 184 Jen Theodore/Unsplash, p. 187 Ilaria de Bona/Unsplash, p. 189 Illustration: Eva Stadler, s. 191 Zuri Swimmer/Alamy/Alamy Stock Photos/mauritius images, p. 192 IanDagnall Computing/Alamy/Alamy Stock Photos/mauritius images, p. 193 IanDagnall Computing/Alamy/Alamy Stock Photos/mauritius images, p. 195 The Picture Art Collection/Alamy/Alamy Stock Photos/mauritius images, p. 197 Vidimages/Alamy/Alamy Stock Photos/mauritius images, p. 199 akg-images/© Georgia O´Keeffe Museum/VG Bild-Kunst, Bonn 2023, p. 201 SuperStock/WilliP/mauritius images/© VG Bild-Kunst, Bonn 2023, p. 202/203 Illustration: Eva Stadler, p. 206/207 Illustration: Eva Stadler

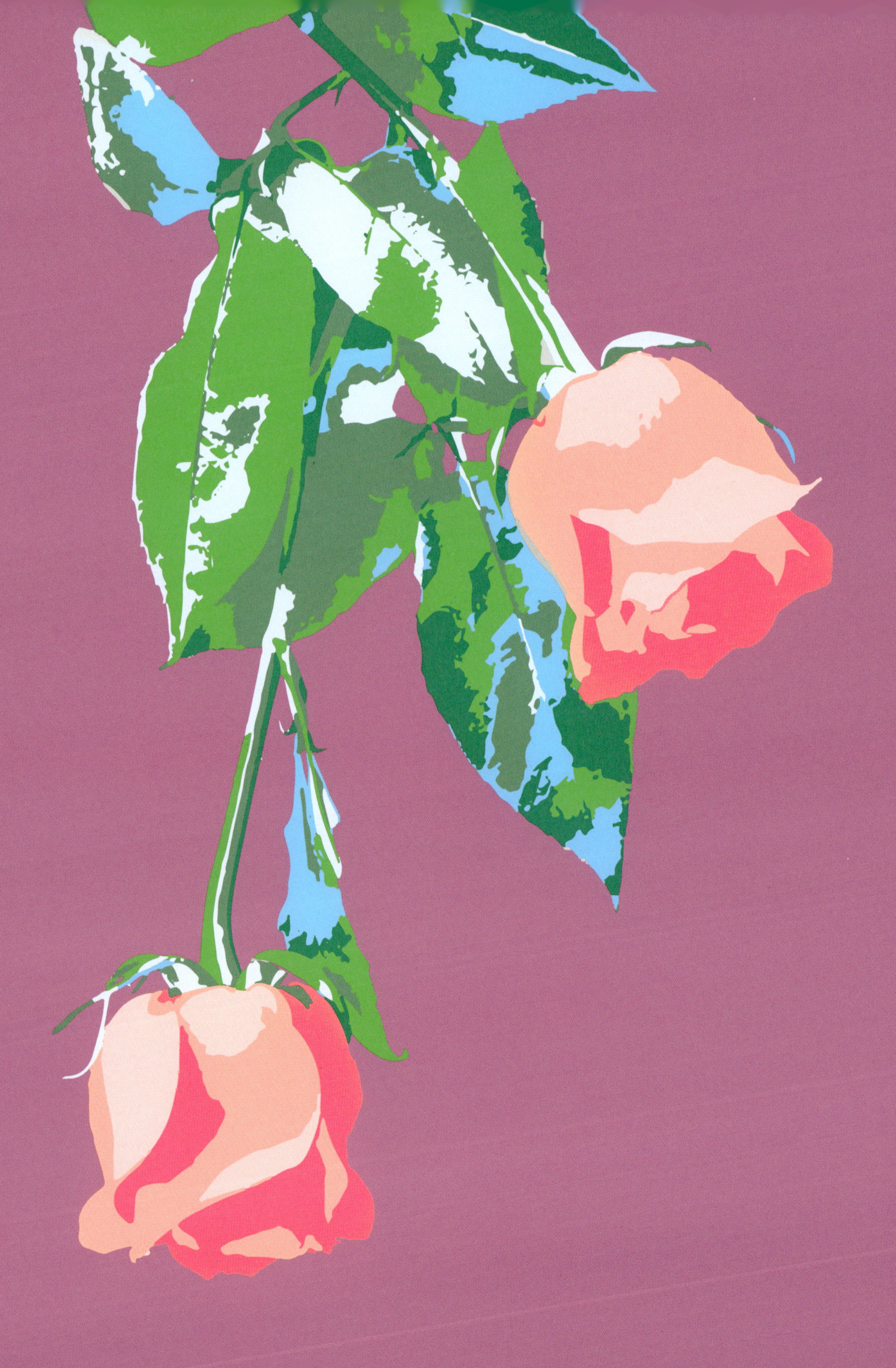

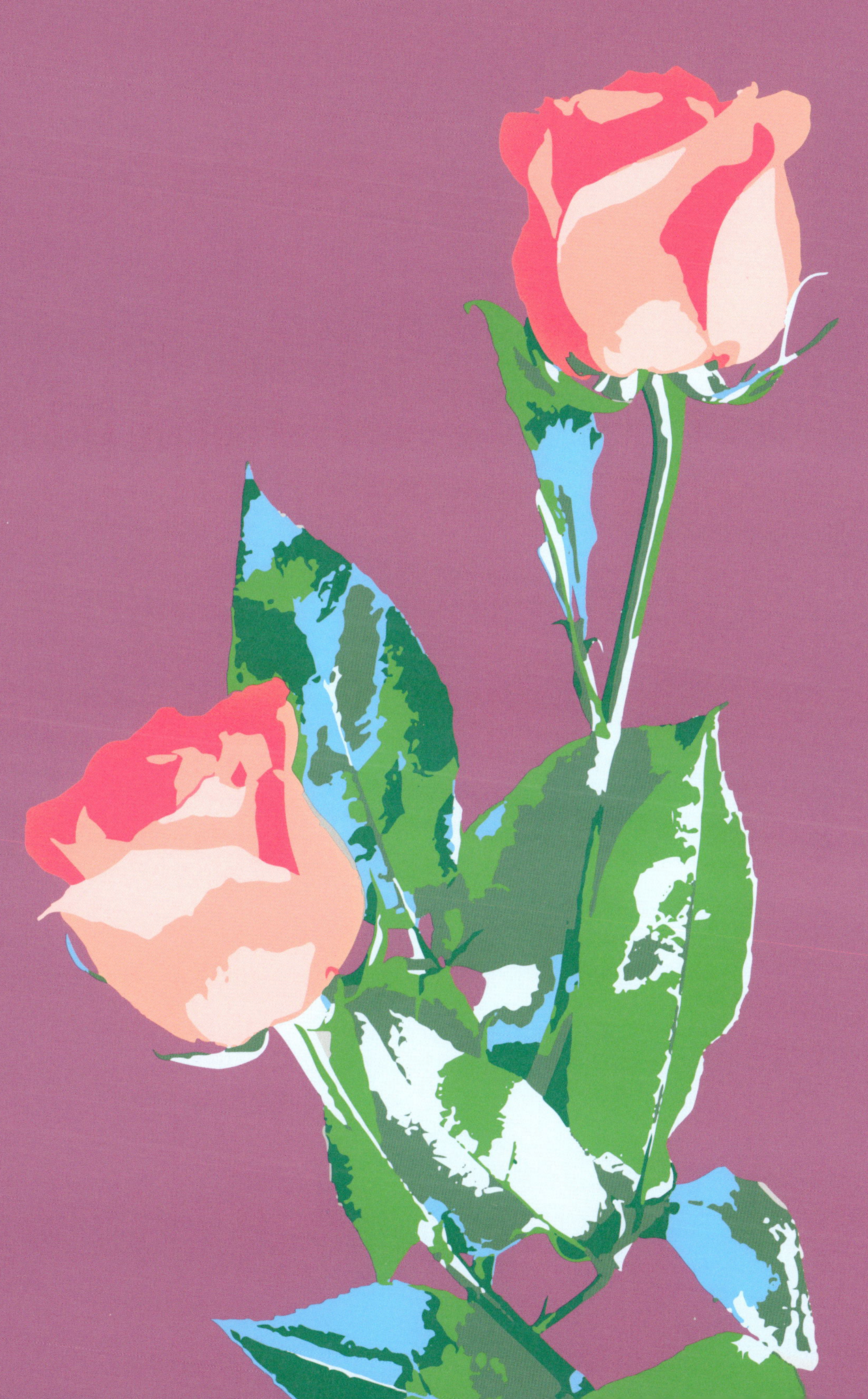

IMPRINT | IMPRESSUM

© 2023 teNeues Verlag GmbH

Text: Anja Klaffenbach
English Translation: John A Foulks
Picture Editing: Heide Christiansen
Design: Eva Stadler
Copy Editing: Dunja Reulein
Proofreading: Simona Fois
Editorial Coordination: Johannes Abdullahi, teNeues Verlag
Production: Sandra Jansen, teNeues Verlag
Photo Editing, Color Separation: Jens Grundei, teNeues Verlag

ISBN: 978-3-96171-481-0
Library of Congress Number: 2023932558
Printed by GPS in BiH

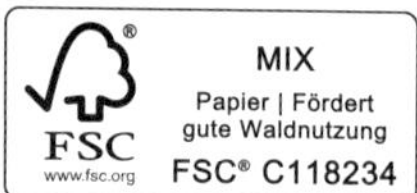

Picture and text rights reserved for all countries. No part of this publication may be reproduced in any manner whatsoever.

While we strive for utmost precision in every detail, we cannot be held responsible for any inaccuracies, neither for any subsequent loss or damage arising.

Every effort has been made by the publisher to contact holders of copyright to obtain permission to reproduce copyrighted material. However, if any permissions have been inadvertently overlooked, teNeues Publishing Group will be pleased to make the necessary and reasonable arrangements at the first opportunity.

Bibliographic information published by the Deutsche Nationalbibliothek:
The Deutsche Nationalbibliothek lists this publication in the Deutsche Nationalbibliografie; detailed bibliographic data are available on the Internet at dnb.dnb.de.

Published by teNeues Publishing Group

teNeues Verlag GmbH
Ohmstraße 8a
86199 Augsburg, Germany

Düsseldorf Office
Waldenburger Straße 13
41564 Kaarst, Germany
e-mail: books@teneues.com

Augsburg/München Office
Ohmstraße 8a
86199 Augsburg, Germany
e-mail: books@teneues.com

Berlin Office
Lietzenburger Straße 53
10719 Berlin, Germany
e-mail: books@teneues.com

Press Department
e-mail: presse@teneues.com

teNeues Publishing Company
350 Seventh Avenue, Suite 301
New York, NY 10001, USA

www.teneues.com

teNeues Publishing Group
Augsburg / München
Berlin
Düsseldorf
London
New York

teNeues

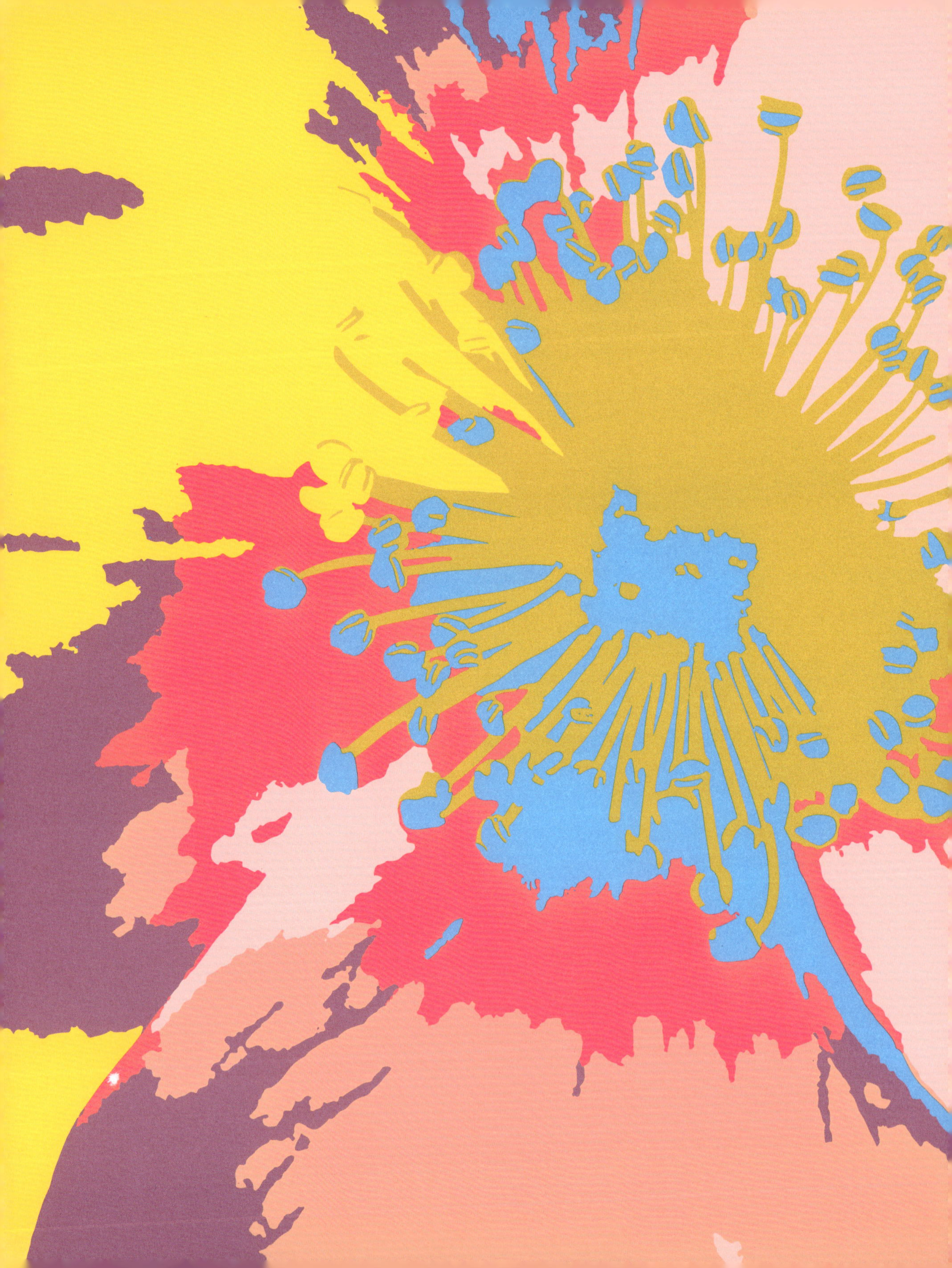